Gas Well Deliquification

Gas Well Deliquification

Editor

Jostna Chitnis

scitus
academics

Gas Well Deliquification

Edited by **Jostna Chitnis**

Printed in 2017

ISBN: 978-1-68117-385-6

Library of Congress Control Number: 2015936535

© 2016 by
SCITUS Academics LLC,
616, Corporate Way, Suite 2, 4766,
Valley Cottage, NY 10989

www.scitusacademics.com

Contents

Preface

Gas well deliqufication, also referred to as "gas well dewatering", is the general term for technologies used to remove water or condensates build-up from producing gas wells. When natural gas flows to the surface in a producing gas well, the gas carries liquids to the surface if the velocity of the gas is high enough. A high gas velocity results in a mist flow pattern in which liquids are finely dispersed in the gas. Consequently, a low volume of liquid is present in the tubing or production conduit, resulting in a pressure drop caused by gravity acting on the flowing fluids. As the gas velocity in the production tubing drops with time, the velocity of the liquids carried by the gas declines even faster. Flow patterns of liquids on the walls of the conduit cause liquid to accumulate in the bottom of the well, which can either slow or stop gas production altogether.

Editor

Chapter 1

A Study of Continuous Flow Gas Lift System Using CFD

Siti Syahida Mohd. Yasin, Nik Mohd. Asraf Nik Aziz,
Zainal Zakaria, and Ariffin Samsuri

Department of Renewable Energy Engineering, Faculty of Petroleum
Engineeringt and Renewable Energy Engineering, Universiti Teknologi
Malaysia, Johor, Malaysia

ABSTRACT

Gas lift optimization has been a point of interest for some decades in order to increase production in oil fields. Despite that there are still lacks of user friendly software in gas lift optimization. The ultimate goal of this study is to develop user friendly program using commercial Computational Fluid Dynamics (CFD) software for optimization of continuous flow gas lift for single well. GAMBIT 2.3.16 was used for modelling the tubing string prior the simulation process. Two-phase CFD calculations using Eulerian–Eulerian model and commercial CFD package FLUENT 6.0 were employed to calculate the gas-liquid flow in the string. The depth and the rate of gas injection were the main parameter tested. The simulation results showed that the highest velocity of flowing fluid was observed when the gas was injected at the lowest point, which is at 5 m from the casing shoe in this study. The velocity of flowing fluid was increased by 1.7% when the gas injection rate increased 20%. This interrelationship was observed with regard of the decrease in liquid column density. CFD is capable in aiding faster continuous flow gas lift optimization process.

INTRODUCTION

Global demand on petroleum was never decrease. Since it is finite and scarce natural resources, petroleum industry players are looking forward for more efficient technologies in all aspects of optimum production. During initial stage of production, the bottom-hole pressure (BHP) in the oil reservoir is sufficient to force the flow of oil to the surface naturally. However, as time goes by, the internal pressure of depleted reservoir can force only a fraction of it. Thus, the use of artificial technique becomes essential.

Continuous flow gas lift being one of the most economic artificial lift method employed in offshore operations worldwide since it is flexible in its production rates, very effective in a wide range of operating condition as it is able to handle corrosive fluids, suitable for high gas oil ratio, high temperature wells, high water cut offshore well (Mahmoud Safar et al., 2011) and compatible with sand production. Other than that, continuous flow gas lift technique required less maintenance compared with other alternatives (Ayatollahi et al., 2004; Naguib et al., 2000).

The main idea in continuous flow gas lift system is that by compressing gas at the surface and injecting it as deep as possible into the well to reduce the density of the well fluid and the hydrostatic pressure loss along the liquid flow path. The reduction of pressure drop in the well caused the pressure in the bottom becomes sufficiently low to continue production for low producing and non-producing well (Mahmoud Safar et al., 2011; Aamo et al., 2005; Szucs and Lim, 2005).

From a production technologist's point of view, if the continuous flow gas lift system has already been installed, the left over question will be focus on the maximum production of oil that could be achieved and the amount of gas required in maintaining the maximum production. In contrast, if the continuous flow gas lift system has not been installed, the problem will be focused on determining the optimal position of the injection point in order to maximize the oil production. Obtaining the optimum gas injection rate is important because although oil production increase as gas injection increased, injection of excessive gas not only will reduce production rate but also increase the operation cost due to high gas prices and compressing costs (Buitrago et al., 1996; De Souza et al., 2010).

Due to this situation, many researches as well as oil operators are working hard to develop a procedure to optimize the lift gas in continuous flow gas lift, both in design and operating phase to produce a well or a group of wells with optimal economic point and at the same time, enhancing oil production (Kanu et al., 1981). In 1981, a method of equal slope allocation under both limited and unlimited gas supply was developed to allocate a total amount of gas at the optimal economic point for a group of wells (Kanu et al., 1981). Fifteen years later, the same results were utilized to establish a new tool to optimize gas distribution for a group of wells without restriction in a well response and the number of wells in the system (Buitrago et al., 1996). As gas lift optimization is an interesting area that can be improved, many researchers explored this problem in determining the optimal operational conditions to yields maximum total oil production rate for single well models (Fang and Lo, 1996) and multiple wells model (Alarcon et al., 2002; Ray and Sarker, 2007; Saepudin et al., 2010), either the gas supply is limited (Saepudin et al., 2010) or unlimited (Khamehchi et al., 2009a) using different formulations such as a linear programming (Fang and Lo, 1996), non-linear programming (Alarcon et al., 2002), mixed integer linear programming (Kosmidis et al., 2005), dynamic programming (Camponogara and Nakashima, 2006), genetic algorithms (Ray and Sarker, 2007; Khamehchi et al., 2009b; Saepudin et al., 2010), Integrated Asset Model (Fernando et al., 2007), Evolutionary Algorithms (Ray and Sarker, 2008), Artificial Neural Networks (Khamehchi et al.,2008) and Ant Colony Approach (Zerafat et al., 2009).

In the recent years, due to progress in computer hardware and software and consequent increased of the calculation speed, Computational Fluid Dynamic (CFD) technique has been powerful and effective tools to understand the complex hydrodynamics of gas liquid two-phase flows and it has been successfully tried and tested in many industries including the oil and gas industry. CFD has been used to model fluid behavior in reservoirs, near wellbores and wells, sand transport, coupled mechanical modeling of failing wellbores and gas well clean-up. Most recently, CFD has mainly been used to model fluid flow in surface pipelines in order to examine the fluid flow through and around tubing and equipment (Alizadehdakhel et al., 2009; Byrne et al., 2011).

Thus, in this study, the CFD modelling was used to propose a framework for the analysis of gas-liquid flow model in continuous flow gas lift system. Variation of injection depth on vertical upward flowing tubing string and the effect of injection gas rate will be considered and discussed.

MATERIALS AND METHODS

Prior to simulation, the tubing string was modelled by GAMBIT 2.3.16 with a few assumptions and boundary condition that will be described detailed in next section. Then, the simulation process was carried out with FLUENT 6.2.16.

Modelling the Tubing String

The model of tubing string for artificial gas lifts was modelled by GAMBIT 2.3.16 with all assumptions and boundary condition being made as described in the next section. The study consists of a well tubing string which crude oil and hydrocarbon gas enters at the specific injection port at certain depth in the domain. The primary phase in the system (crude oil) was entered at certain pressure define as BHP while the injected gas was being well specify by determine the specific rate needed to produce certain volume of oil. Table 1 shows the relevant information and properties of the well while the properties of the crude oil and hydrocarbon gas showed in Table 2.

Based on the ability of computer hardware and time constraints, the whole tubing with actual length of 2000 m is impossible to be modelled as infinite iteration process occurs when the length of the tubing specified as the actual length. This factor alone needs a very high numbers of meshes due to very large difference between length of pipe and its diameter.

Only the lowest 30 meters of the well was modelled with proper assumptions being created. It was assumed that the liquid crude oil and injected gas was well mixed before reaching height of 30 meters. Quadrilateral/hexahedron mesh could be used in tubing till reaching the outlets. However, the section at the gas inlet or specifically at the gas injection section is highly unstructured. Hence, Triangle/Tetrahedron was use in this particular section based on the compatibility of this mesh type. Coarse mesh size of 0.01 m was used in order to have 4480 cells (14134 faces). At the walls, slip boundary condition was imposed. The tubing was modelled as an open system, so the pressure in the outflow space above the initial tubing was set equal to the ambient pressure (0.1 MPa). The development of the model was shown in Fig. 1.

Model Assumptions and Limitations

In order to proceed with the simulation, several assumptions need to be applied in this study based on the software criteria and limitations as stated below:

Siti Syahida Mohd. Yasin, Nik Mohd. Asraf Nik Aziz, Zainal Zakaria, and Ariffin Samsuri

- The atmospheric pressure was 0.1 MPa while the gravitational acceleration was 9.81 m s^{-1}
- Negligible of formation gas in volume and pressure
- Injection gas pressure was higher than the BHP as it was low in most low production wells where gas injection was needed
- The crude oil and hydrocarbon gas entered the tubing were assumed uniformly distributed at the inlet cross section while the mixture of oil and hydrocarbon gas was assumed to be homogenous mixture before reaching the model boundary maximum length at 30 meters. This indicates a higher position of gas injection from well bottom was applied where the flowing

BHP was high enough to drive the oil to a higher location in the tubing

- Radius of gas inlet was set at 1/3 of radius of tubing size since varying the tubing size without careful calculations will lead to divergence in iteration
- The well was packed off in this study due to packerless wells simulations require set up of a porous zone or oil zone outside the well bottom boundary
- Single string of gas injection was used to avoid the complicated multi phase flow and existence of secondary flow

Table 1: Specification of production well

Parameter	Value
Tubing depth(m)	2000
Tubing diameter(m)	0.90
Value port diameter(m)	0.03
Bottom hole pressure(MPa)	5
Gas injection rate(m sec-1)	2.184
Well head pressure(bar)	1
Crude oil daily production(m3day-1)	95.3
Gas-Liquid-ratio (GLR), (m3 gas m-3 oil)	1:1
Actual velocity, Vact(m sec-1)	0.023
Actual Pressur, Pact(KPa)	1
Viscosity of gad, μ(cp)	0.012
Specific gravity of gas, S	0.61

Table 2: Properties of component

Component	Density (kg m-3)	Viscosity(Pa.s)
Liquid oil	1050.00	0.48
Hydocarbon	10.95	0.07×10-4

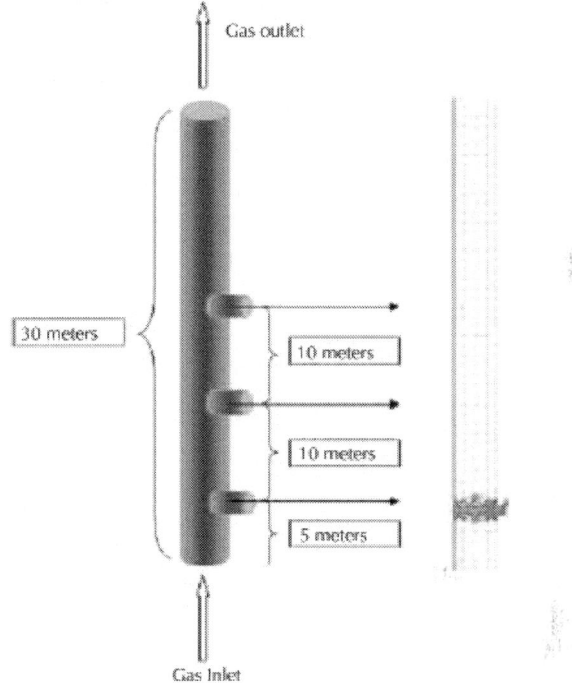

Figure 1: Geometry of tubing string.

Simulation process

The commercial CFD software package, FLUENT 6.2.16, which is based on the 3D segregated 1st order implicit solver approach, was used for solving the set of governing equations for multi phase calculations.

Table 3: various configurations of models used in simulation

No	Hi+j	U1(m sec-1)	Ug(m sec-1)
1	5	2.568	1.747
2	5	2.568	2.184
3	5	2.568	2.184
4	15	2.568	2.184
5	25	2.568	2.184

Using the segregated solver, the flow of multi phase flow of liquid and gas should be accept as in unsteady state. The standard k-ε mixture multi phase model was used to treat turbulence phenomena in both phases. Energy Equation is turned off. Mixture Model was used in this simulation for multi phase flow modelling. The simulation was focus more on the velocity magnitude of the flowing fluids. Both contour and vector of the fluid will be process from the result and the residual plot is turn on during the simulation. The modelling and simulation involves following steps to ensure the result convergence thus valid to be used:

- Generation of suitable grid system
- Conversion of governing equation into algebraic equations
- Selection of discretization schemes
- Formulation of the discretized equation at every grid location
- Formulation of pressure equation
- Development of a suitable iteration scheme for obtaining a final solution

To achieve the objective, the simulation was run with three difference depths of injection rate which were 5 m, 15 and 25 m from casing shoe. After the ideal depth of gas injection was determined, the gas injection rates were then varies at 1.747, 2.184 and 2.621 m sec^{-1} to monitor the effect of the varying gas injection rates in oil production. The liquid velocity of the flowing fluid in the tubing string for this simulation was constant at 2.568 sec^{-1}. Table 3 summarizes the different geometrical configurations used in the present work.

RESULTS AND DISCUSSION

Effect of the Depth of Gas Injection

The efficiency of a continuous flow gas lift system was highly influenced by the depth of the gas injection. The velocity of the liquid phase increase when gas was injected at particular depth. The flowing fluid velocity of the gas-liquid multi phase when 2.184 m sec^{-1} of gas was injected at 5, 10 and 25 m from the casing shoe were given in Fig. 2.

From Fig. 2, it shows that the lowest injection point at 5 m has highest velocity of flowing fluids, which were 3.0181 m sec^{-1} while the higher gas injection point (15 m and 25 m) resulted in 2.9157 m sec^{-1} and 2.8339 m sec^{-1} of velocity of flowing fluids.

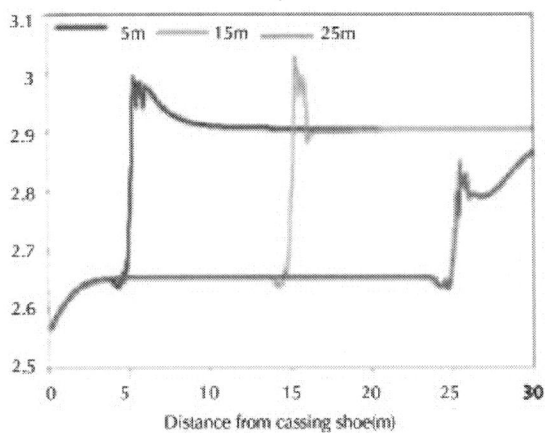

Figure 2: Effect of depth of injection to the oil vertical upward flow.

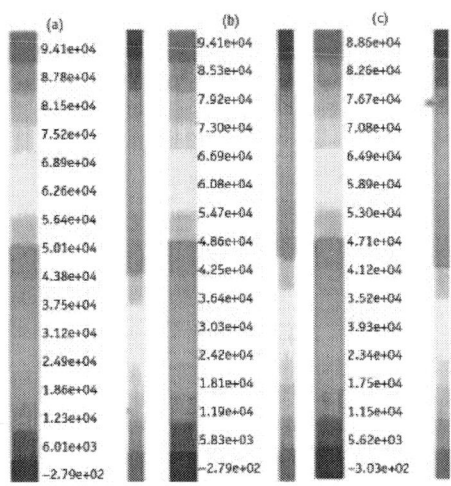

Figure 3(a-c): Contour of hydrostatic pressure for different injection points: (a) 5 m (b) 15 m and (c) 25 m.

It was clearly shows that the higher the gas injection point, the lower the velocity of flowing fluid. When the depth of injection of gas increased, more hydrostatic pressure of the heavier fluid (and gas) column was taken off of the formation thus reducing the BHP and increasing production (Blann and Williams, 1984; Lea et al., 2008). Based on the contour of hydrostatic pressure inside each column (Fig. 3), it was clearly shows that for all columns, the hydrostatic pressure at the bottom was higher when compared to the top of the column. This increment was happened as hydrostatic pressure increased proportionally with depth of wellbore due to the increasing weight of column fluid exerting downward force from the above.

In actual oil field, the gas lift system was installed as deep as possible (Takacs, 2005). The BHP needs to lift the lesser oil volume in the tubing. The deeper the gas injection point, the lower the BHP can be forced. The relationship was not linear but apart from in highly deviated wells, deeper gas lift was not inordinately better. This is because at deeper depths, more reservoir gas was in solution and lift gas therefore has a greater effect. The benefit of injecting gas near the bottom of the most wells can be summarized as (1) More pressure drawdown at the reservoir, (2) Greater total fluid production rates, (3) Less gas volume required, (4) Less downhole equipment and (5) When using a gas injection pressure that enables near-maximum depth injection will unload a well deeper as compared to lower pressure (Blann and Williams, 1984).

Effect of Gas Injection Rates

The quantity of injected gas is a critical variable. The function of injection gas in continuous flow gas lift well is two folds. First, it must aerate the fluid sufficiently to unload the well column down to an operating point. Then, it must reduce the density of the fluid column in order to produce fluid to the surface. However, the fluid production will depends on how much the gas was injected into the column as lower value of injected gas can reduce significantly the production and higher value can increase the operational costs. Figure. 4 shows the effect of gas injection rate at when the gas injection rate varies at 1.747, 2.184 and 2.621 m sec^{-1}. All the gas was injected at 5 m of depth and the velocity of the liquid in the column was constant at 2.568 m sec^{-1}. At minimum rate of gas injection, which was 1.747 m sec^{-1}, the velocity of flowing fluid was 2.8567 m sec^{-1}. The velocity of flowing fluid was keep increasing to 2.90577 and 2.9565 m sec^{-1} when the rate of gas injection was increased to 2.184 and 2.621 m sec^{-1}. Basically, the injected gas will significantly increases the rate of oil production as the gravitational gradient was reduced. The existence of the gas reduced the hydrostatic pressure in the fluid column, in resulted of reduced weight, thus enable the BHP to accelerate the velocities of the flowing liquid vertically upward. This phenomenon should explain the 1.7% increment of the velocity of flowing fluid when the gas injection increased every 20%.

Generally, when a fluid flow through a vertical duct, same as in the wellbore; it will experience pressure losses which can be divided into three different components: 1) Static pressure loss, 2) Dynamic pressure loss and 3) Kinetic pressure loss. For wellbores, the kinetic losses are generally minimal and can be ignored. Thus, the equation that describes the overall pressure losses in the wellbore can be expressed as the sum of two terms:

$$\Delta Ptota; = \Delta Pstatic + \Delta Pdynamic \qquad (1)$$

Eq. 1 is the simplified form of Bernoulli's equation.

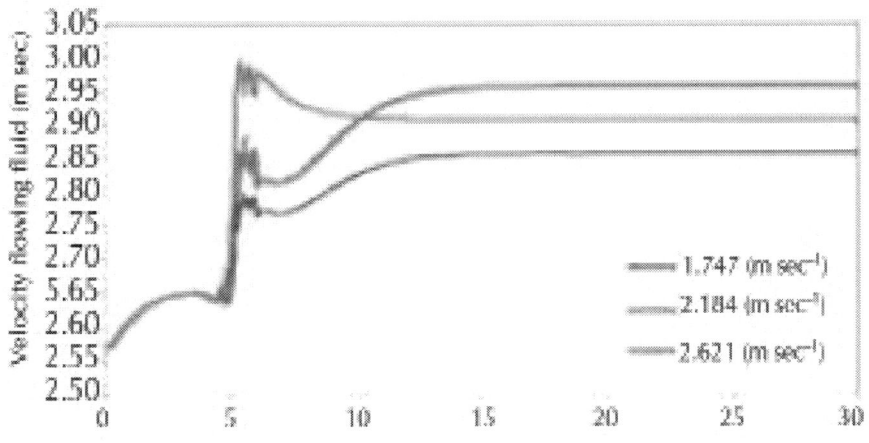

Figure 4: Effect of gas injection rate.

The static pressure losses is a function of the fluid mixture density that exists in the wellbore while the dynamic losses are due to a combination of the particular flow regime that the fluid can be considered to be travel in as well as the composition of the fluid (of gas, liquid and condensate). Thus:

$$\Delta Pstatic = pgh \qquad (2)$$

In gas lift system, when the gravity acceleration, g and the height of the domain is fix at a constant values, h the only terms that can be manipulate is the density, ρ. From the literatures, this is the main cause of introducing gas phase to the flow, where gas phase does actually have lesser density compare to the liquid phase. Thus, it reduces the density of the produced fluid in the tubing. This results in a reduction of the elevational component of the pressure gradient above the point of gas injection and lower down the BHP. Lowering the BHP then increases reservoir drawdown and thus production rate (Kanu et al., 1981).

Such increments have previously been detected for various wells (Fernando et al., 2007; Kanu et al., 1981). However, this effect is eventually offset by increasing frictional losses with increasing injection gas. This is due to in term of the overall pressure gradient, the trade-off to the increased presence of gas is an increased frictional pressure gradient. When the rate of injection gas keeps increasing, it will reach a point, where the benefits of reducing the elevation gradient equal to the drawback of increasing the frictional gradient. In other words, the gas injection has an optimum limit because too much gas injection will cause slippage, where gas phase moves faster than liquid, so that it reduces oil production. Further increase of injection gas will not only has a detrimental effect on the overall production rate, but would not be economically beneficial as it can increase the operational cost with compression and gas usage (De Souza et al., 2010; Fernando et al.,2007). Thus, an economic analysis must be performed to determine the optimum gas injection rate.

CONCLUSIONS

It can be concluded that CFD is capable in aiding faster continuous flow gas lift optimization process as it can be used to determine the correct gas injection depth and the optimum flow rate in gas lift system. Gas lift system is able to increase oil production when the required gas volume is injected through the single valve at the deepest possible depth. The oil production rate was increased by 1.3% when the gas injected into the tubing. The simulation results showed that the highest oil production rate was observed when the gas injected at the lowest point, which is at 5 m from the casing shoe in this study. The oil production rate was increased by 1.7% when the gas injection rate varies from 1.747 m sec^{-1} to 2.621 m sec^{-1}. This interrelationship was observed with regard of the decrease in liquid column density.

ACKNOWLEDGMENTS

Researches would like to thank Ministry of Higher Education Malaysia and Universiti Teknologi Malaysia for the Research University Grant (GUP Project No. Q. J130000.7142.01H93) for their support of this research.

REFERENCES

1. Aamo, O.M., G.O. Eikrem, H.B. Siahaan and B.A. Foss, 2005. Observer design for multiphase flow in vertical pipes with gas-lift-theory and experiments. J. Process Control, 15: 247-257.

2. Alarcon, G.A., C.F. Torres and L.E. Gomez, 2002. Global optimization of gas allocation to a group of wells in artificial lifts using nonlinear constrained programming. J. Energy Resour. Technol., 124: 262-268.

3. Alizadehdakhel, A., M. Rahimi, J. Sanjari and A.A. Alsairafi, 2009. CFD and artificial neural network modeling of two-phase flow pressure drop. Int. Commun. Heat Mass Transfer, 36: 850-856.

4.   Ayatollahi, S., M. Narimani and M. Moshfeghian, 2004. Intermittent gas lift in Aghajari oil field, a mathematical study. J. Petroleum Sci. Eng., 42: 245-255.

5.   Blann, J.R. and J.D. Williams, 1984. Determining the most profitable gas injection pressure for a gas lift installation. J. Petroleum Technol., 36: 1305-1311.

6.   Buitrago, S., E. Rodriguez and D. Espin, 1996. Global optimization techniques in gas allocation for continuous flow gas lift systems. Proceedings of the Conference of Gas Technology, April 28-May 1, 1996, Calgary, Alberta, Canada, pp: 1-9.

7.   Byrne, M., M. Jimenez, E. Rojas and E. Castillo, 2011. Computational fluid dynamics for reservoir and well fluid flow performance modelling. Proceedings of the SPE European Formation Damage Conference, June 7-10, 2011, Noordwijk, The Netherlands, pp: 1-7.

8.   Camponogara, E. and P. Nakashima, 2006. Optimizing gas lift optimization problem using dynamic programming. Eur. J. Oper. Res., 174: 1220-1246.

9.   De Souza, J.N.M., J.L. De Medeiros, A.L.H. Costa and G.C. Nunes, 2010. Modeling, simulation and optimization of continuous gas lift systems for deepwater offshore petroleum production. J. Petroleum Sci. Eng., 72: 277-289.

10. Fang, W.Y. and K.K. Lo, 1996. A generalized well management scheme for reservoir simulation. SPE Reservoir Eng., 11: 116-120.

11. Fernando, G., H. Aron, S. Mack and R. Kashif, 2007. A new approach to gas lift optimization using an integrated asset model. Proceeding of the International Conference on Petroleum Technology, December 4-6, 2007, Dubai, pp: 1-10.

12. Kanu, E., J. Mach and K. Brown, 1981. Economic approach to oil production and gas allocation in continuous gas lift. J. Petroleum Technol., 33: 1887-1892.

13. Khamehchi, E., F. Rashidi, B. Karimi, P. Pourafshary and M. Amiry, 2009. Continuous gas lift optimization using genetic algorithm. Aust. J. Basic Applied Sci., 3: 3919-3929.Khamehchi, E., F. Rashidi, H. Omranpour, S. Shiry Ghidary, A. Ebrahimian and H.

14. Rasouli, 2009. Intelligent system for continuous gas lift operation and design with unlimited gas supply. J. Applied Sci., 10: 1889-1897.

15. Khamehchi, E., F. Rashidi, H. Rasouli, Rasouli and A. Ebrahimian, 2008. A new approach to continuous gas lift optimization using artificial neural networks. J. Eng. Comput. Kosmidis, V.D., J.D. Perkins and E.N. Pistikopoulos, 2005. A mixed integer optimization formulation for the well scheduling problem on petroleum fields. Comp. Chem. Eng., 29: 1523-1541.

16. Lea, J.F., H.V. Nickens and M.R. Wells, 2008. Gas Well Deliquification: Gas Lift. Elsevier Inc., Amsterdam, Netherlands, ISBN-13: 9780080569406, Pages: 608.

17. Mahmoud Safar, B., M. Saedi, S. Mohammad Hossein and A. Sepehr, 2011. Design of a gas lift system to increase oil production from an Iranian offshore well with high water cut. Aust. J. Basic Applied Sci., 5: 1561-1565.

18. Naguib, M.A., S.E. Shaheen, A. El-Wahab Bayoumi and N.A. El-Emam, 2000. Review of artificial lift in Egypt. Proceeding of the Conference and Exhibition on SPE Asia Pacific Oil and Gas, October 16-18, 2000, Brisbane, Australia, pp: 1-9

19. Ray, T. and R. Sarker, 2007. Genetic algorithm for solving a gas lift optimization problems. J. Pet. Sci. Eng., 59: 84-96.

20. Ray, T. and R. Sarker, 2008. Evolutionary algorithms deliver promising results to gas lift optimization problems. World Oil, 229: 141-142.

21. Saepudin, D., P. Sukarno, E. Soewono, K.A. Sidarto and A.Y. Gunawan, 2010. Oil production optimization in a cluster of gas lift wells system. J. Applied Sci., 10: 1705-1713.

22. Szucs, A. and F. Lim, 2005. Heavy oil gas lift using the concentric offset riser (COR). Proceeding of the SPE International Thermal Operations and Heavy Oil Symposium, November 1-3, 2005, Alberta, Canada, pp: 1-5.

23. Takacs, G., 2005. Gas Lift: Manual. 1st Edn., PennWell Books, Tulsa, OK., USA., ISBN-13: 9780878148059, Pages: 478.

24. Zerafat, M.M., S. Ayatollahi and A.A. Roosta, 2009. Genetic algorithm and ant colony approach for gas-lift allocation optimization. J. Jpn. Petroleum Inst., 52: 102-107.

The Contamination of Commercial 15N2 Gas Stocks with 15N–Labeled Nitrate and Ammonium and Consequences for Nitrogen Fixation Measurements

Richard Dabundo[1], Moritz F. Lehmann[2], Lija Treibergs[1], Craig R. Tobias[1], Mark A. Altabet[3], Pia H. Moisander[4], and Julie Granger[1]

[1]University of Connecticut Avery Point, Department of Marine Sciences, Groton, CT, United States of America

[2]University of Basel, Institute for Environmental Geosciences, Basel, Switzerland

[3]University of Massachusetts Dartmouth, Department of Estuarine and Ocean Sciences, North Dartmouth, MA, United States of America

[4]University of Massachusetts Dartmouth, Department of Biology, North Dartmouth, MA, United States of America

ABSTRACT

We report on the contamination of commercial 15-nitrogen (^{15}N) N_2 gas stocks with ^{15}N-enriched ammonium, nitrate and/or nitrite, and nitrous oxide. $^{15}N_2$ gas is used to estimate N_2 fixation rates from incubations of environmental samples by monitoring the incorporation of isotopically labeled $^{15}N_2$ into organic matter. However, the microbial assimilation of bioavailable ^{15}N-labeled N_2 gas contaminants, nitrate, nitrite, and ammonium, is liable to lead to the inflation or false detection of N_2 fixation rates. $^{15}N_2$ gas procured from three major suppliers was analyzed for the presence of these ^{15}N-contaminants. Substantial concentrations of ^{15}N-contaminants were detected in four Sigma-Aldrich $^{15}N_2$ lecture bottles from two discrete batch syntheses. Per mole of $^{15}N_2$ gas, 34 to 1900 µmoles of ^{15}N-ammonium, 1.8 to 420 µmoles of ^{15}N-nitrate/nitrite, and ≥21 µmoles of ^{15}N-nitrous oxide were detected. One $^{15}N_2$ lecture bottle from Campro Scientific contained ≥11 µmoles of ^{15}N-nitrous oxide per mole of $^{15}N_2$ gas, and no detected ^{15}N-nitrate/nitrite at the given experimental $^{15}N_2$ tracer dilutions. Two Cambridge Isotopes lecture bottles from discrete batch syntheses contained ≥0.81 µmoles ^{15}N-nitrous oxide per mole $^{15}N_2$, and trace concentrations of ^{15}N-ammonium and ^{15}N-nitrate/nitrite. $^{15}N_2$ gas equilibrated cultures of the green algae *Dunaliella tertiolecta* confirmed that the ^{15}N-contaminants are assimilable. A finite-differencing model parameterized using oceanic field conditions typical of N_2 fixation assays suggests that the degree of detected ^{15}N-ammonium contamination could yield inferred N_2 fixation rates ranging from undetectable, <0.01 nmoles N L^{-1} d^{-1}, to 530 nmoles N L^{-1} d^{-1}, contingent on experimental conditions. These rates are comparable to, or greater than, N_2 fixation rates commonly detected in field assays. These results indicate that past reports of N_2 fixation should be interpreted with caution, and demonstrate that the purity of commercial $^{15}N_2$ gas must be ensured prior to use in future N_2 fixation rate determinations.

INTRODUCTION

Nitrogen (N) is a major nutrient required universally by photosynthetic organisms. Its availability in the environment can directly affect the ecology and productivity of terrestrial and marine ecosystems, with important implications for the regional and global carbon cycles. The natural input of bioavailable N to the biosphere is dominated by nitrogen fixation, the biological reduction of dinitrogen (N_2) gas to ammonium (NH_4^+). Two methods are commonly utilized to measure N_2 fixation rates in the field, the $^{15}N_2$ tracer assay [1] and the acetylene (C_2H_2) reduction assay [2, 3]. The $^{15}N_2$ tracer assay was originally developed when artificially ^{15}N-enriched substrate N_2 first became available [4]. This approach was then superseded by the acetylene reduction technique, as the cost and availability of high precision isotope ratio measurements proved restrictive [3]. The acetylene reduction assay, however, is associated with variations in the factor used to convert C_2H_2 reduction into N_2 equivalents, and with potentially biasing effects of C_2H_2 on the physiology of N_2 fixing organisms, among other issues [5]–[7]. Interest in the $^{15}N_2$ tracer assay later regained momentum, owing to the increased affordability of Isotope Ratio Mass Spectrometry (IRMS) instrumentation and to concurrent developments in ^{15}N tracer techniques. Today, it is generally the preferred method to quantify N_2 fixation rates in both terrestrial and aquatic environments [1], owing to its high sensitivity, and ability to provide qualitative and quantitative constraints on the translocation and the fate of biologically fixed N [8]–[10].

A salient strength of the $^{15}N_2$ tracer assay is that ^{15}N-enrichment detected in biomass can be ascribed to the biological reduction of N_2 exclusively, as no interfering processes can carry out the reduction of $^{15}N_2$ gas concurrently. This premise requires that the $^{15}N_2$ stock be devoid of any contaminant ^{15}N-species that could be assimilated into biomass simultaneously. However, during recent research projects on N_2 fixation conducted independently in our laboratories at the University of Connecticut Avery Point and the University of Massachusetts Dartmouth, convergent observations indicated that some commercial $^{15}N_2$ stocks could be contaminated with ^{15}N-enriched N-species other than N_2, including nitrate, nitrite and/or ammonium. These reactive forms of N would be readily assimilated by microorganisms, leading to significantly biased (i.e., overestimated) N_2 fixation rate measurements. These observations motivated the current study, with the goal of testing whether commercially available $^{15}N_2$ stocks contain ^{15}N-contaminants at levels that would interfere with $^{15}N_2$ tracer N_2 fixation assays, particularly in the open ocean, and to assess if such contaminants are prevalent among $^{15}N_2$ stocks from different suppliers.

We thus uncovered substantial contamination of one of three brands of commercial $^{15}N_2$ gas with bioavailable inorganic ^{15}N-species. Our findings raise important concerns regarding the pervasiveness of reactive ^{15}N contamination of the $^{15}N_2$ stocks, and the extent to which these contaminants may have affected the magnitude of the N_2 fixation rate estimates reported in the literature. We outline steps to contend with this issue to ensure the veracity of future N_2 fixation estimates.

METHODS

Reagents

Four 33 mL lecture bottles of 98+ at% ^{15}N-labeled N_2 gas were purchased from Sigma-Aldrich (produced by their subsidiary, Isotec Stable Isotopes; St. Louis, MO; Stock Keeping Unit 364584), three from lot # SZ1670V, synthesized in 2010, and one from lot # MBBB0968V, synthesized in 2014. Two 1L lecture bottles of 98+ at% $^{15}N_2$ were purchased from Cambridge Isotopes (Tewksbury, MA, part # NLM-363-1-LB) from respective lot #'s I1-11785A and I-16727. One 1L lecture bottle of 98+ at% $^{15}N_2$ was purchased from Campro Scientific (Berlin, Germany; catalogue # CS01-185_261) from lot # EB1169V. Ammonium and nitrate solutions were prepared with salts or with solutions obtained from different distributors: sodium nitrate ($NaNO_3$: CAS 7631-99-4), potassium nitrate (KNO_3: CAS 7757-79-1), and ammonium chloride (NH_4Cl: CAS 12125-02-9) from Fisher Scientific (Pittsburgh, PA); analytical-grade potassium nitrate (CAS 7757-79-1) from Fluka Analytical and a gravimetric solution of ammonium chloride (catalogue # AS-NH3N9-2Y) from SPEX CertiPrep (Metuchen, NJ).

Preparation of Nitrate and Ammonium Solutions Equilibrated with $^{15}N_2$ gas

In order to determine whether the $^{15}N_2$ gas stocks contained ^{15}N-labeled ammonia (NH_3) or nitrate and/or nitrite (NO_x) contaminants, aqueous solutions of natural abundance (unlabeled) ammonium and nitrate salts were equilibrated overnight with an air headspace supplemented with an injection of $^{15}N_2$ gas. After equilibration, the $^{15}N/^{14}N$ ratio of ammonium and the $^{15}N/^{14}N$ and $^{18}O/^{16}O$ ratios of nitrate/nitrite in solution were measured, as well as the $^{15}N/^{14}N$ ratio of N_2 gas in the headspace, as described below. The isotope ratios of nitrate and ammonium were compared to those in control solutions, which were not supplemented with $^{15}N_2$ gas. Experiments with the Campro Scientific $^{15}N_2$ stock were verified for ^{15}N-nitrate/nitrite contaminants only (and not for ^{15}N-ammonium).

Initial experiments consisted of 40 mL or 100 mL solutions of 10, 50, 100, 200, or 300 µmol L^{-1} nitrate and 5 µmol L^{-1} ammonium chloride in 60 mL or 120 mL serum vials that were sealed with Thermo Scientific gas-impermeant stoppers (part # C4020-30) or with Bellco Glass septum stoppers (catalogue # 2048-11800). The 20 mL of air headspace in each of the treatment vials was supplemented with 0.1 mL of $^{15}N_2$ gas from respective bottles from each of the three suppliers (three lecture bottles from Sigma-Aldrich lot # SZ1670V and one bottle from lot # MBBB0968V, two bottles from Cambridge Isotopes lot # I1-11785A and lot # I-16727, and one bottle from Campro Scientific lot # EB1169V). The solutions were equilibrated overnight on a shaker, after which the $^{15}N/^{14}N$ and $^{18}O/^{16}O$ isotope ratios of nitrate were analyzed as described below. The $^{15}N/^{14}N$ isotope ratio of ammonium was also analyzed (described below) in experimental solutions treated with the Sigma-Aldrich and Cambridge Isotopes stocks, but not the Campro Scientific stock.

The experimental sensitivity to ^{15}N-contaminants was increased in subsequent experiments involving $^{15}N_2$ stocks that did not show clear evidence of contamination in the experiments described above (see *Results*) by increasing the volume of $^{15}N_2$ gas injections and decreasing solution volumes. Experiments were initiated in which 2 mL $^{15}N_2$ gas was equilibrated overnight in 20 mL serum vials containing 10 mL solutions of 10 µmol L^{-1} sodium nitrate, after which the $^{15}N/^{14}N$ and $^{18}O/^{16}O$ ratios of nitrate were measured as described below. Similarly, 10 mL solutions of 5 µmol L^{-1} ammonium chloride were dispensed in 20 mL serum vials and equilibrated overnight with 2 mL $^{15}N_2$ gas, after which the $^{15}N/^{14}N$ isotope ratios of ammonium were analyzed (described below).

The measured $^{18}O/^{16}O$ ratios of nitrate/nitrite in solutions equilibrated with $^{15}N_2$ gas from some stocks suggested the presence of $^{46}N_2O$ contamination. As our analyte for isotope ratio analysis is N_2O, and $^{46}N_2O$ can be explained by both $^{15}N^{15}N^{16}O$ and $^{14}N^{14}N^{18}O$, N_2O that is doubly labeled with ^{15}N is falsely detected as $^{18}O_{NO3}$ enrichment. The presence of $^{46}N_2O$ contamination in $^{15}N_2$ gas was verified directly for one of the Sigma-Aldrich stocks (Lot # SZ1670V) by adding 0.0125, 0.020, or 0.025 mL of $^{15}N_2$ stock to 20 mL serum vials containing 10 nmoles of reference N_2O in helium. The N and O isotopic composition of the N_2O was analyzed as described below, and compared to unamended N_2O injections.

Dunaliella Tertiolecta Cultures

The marine green alga *Dunaliella tertiolecta* was cultured in growth media equilibrated with$^{15}N_2$ gas in order to ascertain the susceptibility of ^{15}N-labeled gas contaminants to assimilation by non-N_2-fixing organisms. Culture medium was prepared from filtered Long Island Sound sea water supplemented with 50 µmol L^{-1} $NaNO_3$, 36.3 µmol L^{-1} $NaH_2PO_4*H_2O$, and 107 µmol L^{-1} $Na_2SiO_3*9H_2O$, as well as f/2 trace metals and f/2 vitamins [11], added from filter sterilized stock solutions. Medium (200 mL) was dispensed in 250 mL stoppered glass bottles. Experimental treatment bottles were equilibrated overnight with 0.2 mL $^{15}N_2$ gas from either a Cambridge Isotopes (lot #I-16727) or Sigma-Aldrich (lot # SZ1670V) lecture bottle. Following inoculation, cultures were left loosely capped and placed on a windowsill with exposure to natural light. Nitrate concentrations were monitored daily. Upon the complete depletion of nitrate, 8 days after inoculation, the cultures were harvested on pre-combusted GF/F filters. Filters were dried at 60°C for 18 h pending N isotopic analysis of the particulate nitrogen (described below).

Nitrate and Ammonium Concentrations

Nitrate concentrations in the experimental solutions were verified *via* reduction to nitric oxide in hot vanadium (III) solution followed by detection with a chemiluminescence NO_x analyzer (model T200 Teledyne Advanced Pollution Instrumentation) [12]. Ammonium concentrations were measured by derivatization with orthophthaldialdehyde (OPA) and fluorometric detection on an AJN Scientific f-2500 Fluorescence Spectrophotometer [13].

Nitrate N and O Isotope Ratio Analyses

Nitrate/nitrite nitrogen ($^{15}N/^{14}N$) and oxygen ($^{18}O/^{16}O$) isotope ratios were measured using the denitrifier method [14,15]. Nitrate (and nitrite) in experimental samples was converted stoichiometrically to nitrous oxide (N_2O) by a denitrifying bacterial strain (*Pseudomonas chlororaphis* f. sp. *aureofaciens*, ATCC 13985) that lacks nitrous oxide reductase. The N and O isotopic composition of N_2O was then measured on a Delta V Advantage Isotope Ratio Mass Spectrometer (IRMS) interfaced with a modified Gas Bench II gas chromatograph (Thermo Fisher) purge and trap system. The isotope ratio measurements are reported in the conventional delta ( ) notation in per mille (‰) units, defined for N and O by the following equations:

$$\delta^{15}N_{sample} =$$

$$\left[\left(^{15}N_{sample}/^{14}N_{sample} \right) \div \left(^{15}N_{reference}/^{14}N_{reference} \right) - 1 \right] * 1,000$$

$$\delta^{18}O_{sample} =$$

$$\left[\left(^{18}O_{sample}/^{16}O_{sample} \right) \div \left(^{18}O_{reference}/^{16}O_{reference} \right) - 1 \right] * 1,000$$

The $^{15}N/^{14}N$ reference is N_2 in air, and the $^{18}O/^{16}O$ reference is Vienna Standard Mean Ocean water (V-SMOW). Individual analyses on the GC-IRMS were referenced to injections of N_2O from a pure N_2O gas cylinder, and then standardized through comparison to the international nitrate standards USGS-34 ($\delta^{15}N$ of −1.8‰ vs. air; $\delta^{18}O$ of −27.9‰ vs. V-SMOW), USGS-32 ($\delta^{15}N$ of +180‰ vs. air; $\delta^{18}O$ of +25.7‰ vs. V-SMOW), and IAEA-NO-3 ($\delta^{15}N$ of +4.7‰ vs. air; $\delta^{18}O$ of +25.6‰ vs. V-SMOW) [16]–[18], using standard bracketing techniques. Nitrate samples from experiments with Campro Scientific $^{15}N_2$ were standardized with USGS-32 and IAEA-NO-3, and an additional internal lab nitrate standard (UBN-1; $\delta^{15}N$ of 14.15‰ vs. air; $\delta^{18}O$ of +25.7‰ vs. V-SMOW). Precision for analytical replicates was ≤0.2‰ for $\delta^{15}N_{NO3}$ and ≤0.2‰ for $\delta^{18}O_{NO3}$ for isotope ratio amplitudes encompassed by the standards. Above 200‰, the precision decreased in proportion to

$^{15}N_{NO3}$ amplitude, with standard deviations of 7.5‰ for $\delta^{15}N_{NO3}$ at or above 1000‰. Similarly, precision decreased with increasing $\delta^{18}O_{NO3}$, with standard deviations of 1.3‰ to 6.1‰ for $\delta^{18}O_{NO3}$ values ≥80‰. The poor precision of the higher range measurements is likely due to the variable contribution of a trace NO_x contaminant in denitrifier preparations with $\delta^{15}N$ and $\delta^{18}O$ values that are in the range of natural abundance samples [19].

Nitrous Oxide N and O Isotope Ratio Analyses

N_2O isotope ratios were measured directly on the GC-IRMS, and referenced against the N_2O tank, which was standardized indirectly by comparison to the $\delta^{15}N$ and $\delta^{18}O$ of nitrate standards.

Ammonium N Isotope Ratio Analyses

The ammonium $\delta^{15}N_{NH4}$ was measured using the hypobromite-azide method [20]. Ammonium in basic solution was converted to N_2O via oxidation to nitrite (NO_2^-) with hypobromite, followed by reduction of nitrite to N_2O with sodium azide in acetic acid. The $\delta^{15}N$ of the N_2O analyte was measured on the GC-IRMS, as outlined above. Measurements were calibrated using solutions made from the international standard ammonium salts, IAEA-N1 and IAEA-N2, with assigned $\delta^{15}N$ values of +0.4‰, +20.3‰ vs. air, respectively [16,17,21,22]. Our standard error for analytical replicates was ≤0.6‰ at relatively low ^{15}N-abundances, but increased substantially for $\delta^{15}N_{NH4}$ from 100‰ to 9000‰, varying from 2.9‰ to as high as 59.7‰. As with the nitrate analyses, the low precision of higher range measurements likely stems from the variable contribution of a trace ammonium or nitrite contaminant with a natural abundance $\delta^{15}N$ value, inadvertently introduced during the analyses.

Headspace N_2 Isotope Ratio Analyses

To measure the $\delta^{15}N$ of N_2 gas in the headspace of experimental samples, 75 µL of headspace was injected into 12 mL Exetainer vials previously flushed with helium, then analyzed on a Gas Bench II GC-IRMS (Delta V Advantage Plus) operated in continuous flow mode. N_2 and (O_2+ Ar) were separated on a 5-Å mole-sieve capillary gas chromatography column. The analyses were standardized with parallel analyses of ambient N_2 gas in air. These direct N_2 gas measurements were carried out for experiments conducted using two of three lecture bottles from Sigma-Aldrich lot # SZ1670V, and for experiments conducted using the lecture bottle from Cambridge Isotopes lot # I1–11785A. The $^{15}N_2$ concentration in the headspace of other experiments was estimated from the tracer injection volume rather than from direct measurements.

Particulate Nitrogen Isotope Ratio Analyses

The $\delta^{15}N$ of particulate nitrogen (PN) was analyzed using a Costech Instruments elemental combustion system (model 4010) coupled to a Thermo Scientific Delta V Advantage IRMS. Analyses were standardized using L-glutamic acid reference materials, USGS-40 ($\delta^{15}N$ of −4.52‰ vs. air) and USGS-41 ($\delta^{15}N$ of +47.57‰ vs. air) [23].

RESULTS

Nitrate solutions equilibrated with any of three $^{15}N_2$ gas stocks from Sigma-Aldrich lot # SZ1670V (referred to hereafter as *'Sigma A1, A2 and A3'*) showed a substantial increase in the $\delta^{15}N$ of nitrate (and possibly nitrite) compared to control solutions in the lower sensitivity nitrate dilutions (Fig. 1a). Respective ^{15}N enrichments evidenced by the $\delta^{15}N_{NO3+NO2}$ were inversely proportional to the concentration of nitrate in the solutions, from nearly 1000‰ at 10 µmol L^{-1} nitrate to 30‰ at 290 µmol L^{-1} nitrate, compared to a $\delta^{15}N_{NO3}$ of 23.5±0.5‰ in the corresponding potassium nitrate control solutions. The ^{15}N enrichments imparted on the nitrate solutions were comparable among the three lecture bottles from this lot (# SZ1670V). The $\delta^{15}N_{NO3+NO2}$ resulting from equilibration with a single Sigma-Aldrich gas stock from lot # MBBB0968V (*'Sigma B'*) was relatively modest, but still significant, averaging 28.4±0.3‰ at 10 µmol L^{-1} nitrate, compared to 23.5±0.5‰ in the corresponding control solutions (Fig. 1a). When tested at the more sensitive experimental dilution, equilibrations of nitrate solutions with the *Sigma B* stock resulted in a $\delta^{15}N_{NO3+NO2}$ of 200.2±70.9‰, compared to a $\delta^{15}N_{NO3}$ of 1.3±0.1‰ in control solutions (Fig. 2a). These measurements thus indicate that N_2 gas stocks sourced from lot # SZ1670V contained 410±80 µmoles of ^{15}N-nitrate and/or nitrite per mole of $^{15}N_2$, whereas the bottle from lot # MBBB0968V contributed 1.8±0.6 µmoles of ^{15}N-nitrate and/or nitrite per mole of $^{15}N_2$ (Table 1). The ^{15}N-nitrate additions were determined by a mass balance calculation:

$$^{15}NO_3{}^- + {}^{15}NO_{2\,added}^- =$$
$$\left[\left(\delta^{15}N_{NO3+NO2,final} * NO_3{}^-_{final} \right) - \left(\delta^{15}N_{NO3,initial} * NO_3{}^-_{initial} \right) \right]$$
$$\div \, \delta^{15}N_{NO3+NO2,added}$$

where $^{15}NO_3{}^- + NO_{2\,added}^-$ is the moles of ^{15}N-labeled nitrate and/or nitrite added by the $^{15}N_2$ gas injection, $\delta^{15}N_{NO3+NO2,added}$ is presumed to be equivalent to the $\delta^{15}N$ of $^{15}N_2$ tracer gas (266,540‰), $NO_3{}^-_{initial}$ and $NO_3{}^-_{final}$ refer to the moles of nitrate in solution before and after $^{15}N_2$ equilibration, and $\delta^{15}N_{NO3'initial}$ and $\delta^{15}N_{NO3+NO2'final}$ refer to the $\delta^{15}N_{NO3}$ of nitrate solutions before and after $^{15}N_2$ equilibration. The quantity of $^{15}N_2$ gas added to experimental treatments was measured explicitly in *Sigma A1* and *A2* bottles, and was calculated from the $^{15}N_2$ injection volumes for experiments treated with *Sigma A3* and *B* stocks.

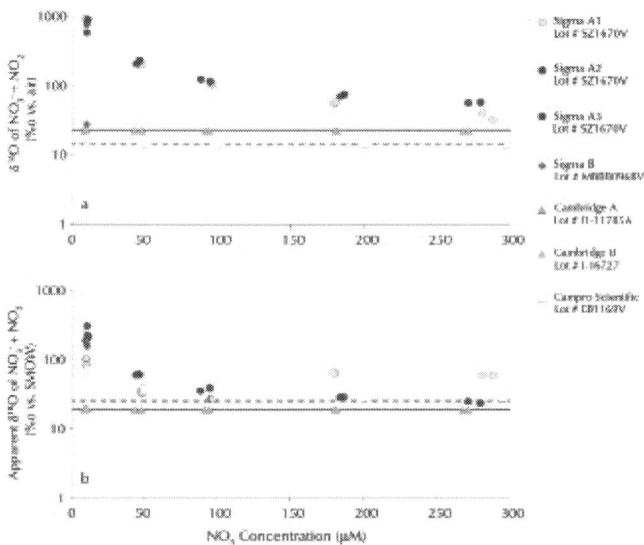

Figure 1: (a) $\delta^{15}N_{NO3+NO2}$ (log scale) of nitrate solutions (10–300 µmol L^{-1}) following equilibration with 0.1 mL $^{15}N_2$ gas from lecture bottles procured

from three distributors. Solutions were 40 mL for Sigma-Aldrich and Campro Scientific equilibrations, and 100 mL for Cambridge Isotopes equilibrations. The solid line corresponds to the $\delta^{15}N_{NO3}$ of the control solutions for Sigma-Aldrich and Cambridge Isotopes experiments ($\delta^{15}N_{NO3} = 23.5 \pm 0.5$‰); the dashed line corresponds to controls for Campro Scientific experiments ($\delta^{15}N_{NO3} = 14.15 \pm 0.1$‰). Paired symbols identify replicate experimental treatments. (**b**) Corresponding apparent $\delta^{18}O_{NO3+NO2}$ of the experimental nitrate solutions. The solid line corresponds to the $\delta^{18}O_{NO3}$ of control solutions for the Sigma-Aldrich and Cambridge Isotope experiments ($\delta^{18}O_{NO3} = 18.9 \pm 0.3$‰); the dashed line corresponds to controls for Campro Scientific experiments (25.4 ± 0.3‰).

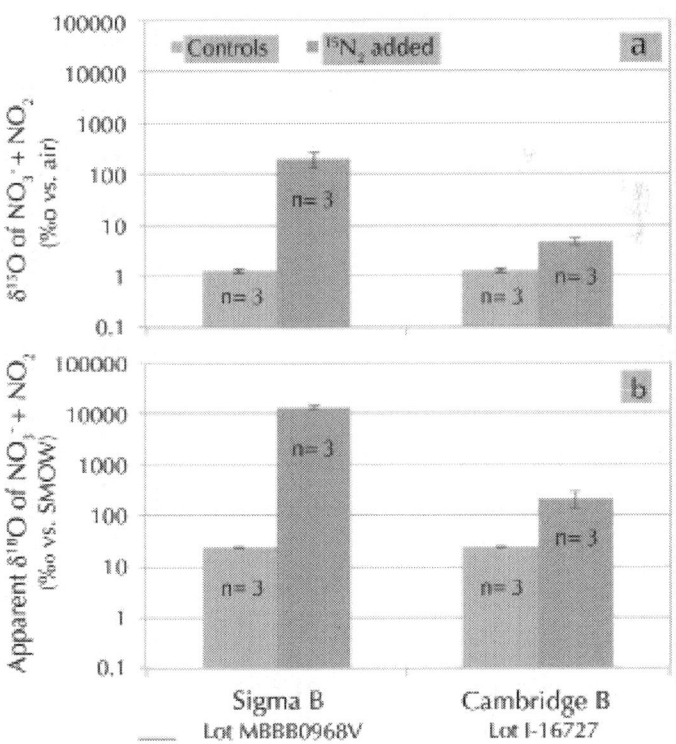

Figure 2: (a) $\delta^{15}N_{NO3+NO2}$ (log scale) of higher sensitivity equilibrations of 10 µmol L^{-1} nitrate solutions (10 mL) with 2 mL of $^{15}N_2$ gas from a Cambridge Isotopes or a Sigma-Aldrich bottle. (b) Corresponding apparent $\delta^{18}O_{NO3}$ (log scale) of higher sensitivity equilibrations of the two stocks. n = the number of experimental replicates.

Table 1: The quantity of ^{15}N-labeled contaminants detected relative to $^{15}N_2$ additions

μmoles 15NX per mole 15N2			
	15NO3–/NO2–	15NH4+	$^{46}N_2O$
Sigma A1 lot # SZ1670V	420±110	34±11	≥21±3
Sigma A2 lot # SZ1670V	420±40	520±30	09±51
Sigma A3 † lot # SZ1670V	350±80	N/A	≥63±15
Sigma B † lot # MBBB0968V	1.8±0.6	1900±560	≥49±17
Cambridge A lot # I1-11785A	n.d.*	0.052±0.020	n.d.*
Cambridge B † lot # I-16727	0.024±0.006	0.014±0.004	≥0.81±0.24
Campro Scientific † lot # EB1169V	n.d.*	N/A	≥11±3

The μmoles of ^{15}N contaminants ($NO_3^-+NO_2^-$, NH_4^+, and N_2O) detected per mole of $^{15}N_2$ gas from lecture bottles provided by different suppliers. N/A = not available; n.d. = not detected.

*Not explicitly tested in high sensitivity $^{15}N2$ dilutions.

'Moles of $^{15}N_2$ estimated from the injection volume rather than direct measurements.
§$^{46}N20$ measured directly.

doi:10.1371/journal.pone.0110335.t0011

In contrast to the Sigma-Aldrich stocks, $^{15}N_{NO3+NO2}$ contaminants were significantly lower, or possibly absent, in Cambridge Isotopes and Campro Scientific $^{15}N_2$ stocks. The $\delta^{15}N_{NO3+NO2}$ values of solutions treated with Cambridge Isotopes $^{15}N_2$ gas (lots # I1–11785A and

I-16727, hereafter referred to as *Cambridge A* and *Cambridge B*, respectively) and with Campro Scientific $^{15}N_2$ gas (lot # EB1169V) were indistinguishable from those of control solutions at all experimental nitrate concentrations in the lower sensitivity tests (Fig. 1a). In the more sensitive experimental treatments, however, solutions treated with *Cambridge B* $^{15}N_2$ gas (lot # I-16727) had a $\delta^{15}N_{NO3+NO2}$ of 4.8±0.8‰, compared to a $\delta^{15}N_{NO3}$ of 1.3±0.1‰ in control solutions (Fig. 2a). This stock thus contributed trace contaminants on the order of 0.024±0.006 µmoles of ^{15}N-nitrate and/or nitrite per mole of $^{15}N_2$ (Table 1). Nitrate isotope ratios in the *Cambridge A* (lot # I1–11785A) and Campro Scientific $^{15}N_2$ gas stocks were not tested at these lower experimental dilutions.

In treatments using several $^{15}N_2$ gas stocks, $\delta^{18}O_{NO3}$ was found to be elevated relative to control solutions. The denitrifier method, employed for $\delta^{18}O_{NO3}$ measurements, involves the bacterial reduction of NO_3^- and NO_2^- to N_2O, and the subsequent analysis of N_2O using an IRMS. However, the elevated $\delta^{18}O_{NO3}$ values detected within experimental treatments are expressly *not* explained by the formation of $^{14}N^{14}N^{18}O$ during bacterial reduction of ^{15}N-enriched nitrate, which could only account for a negligible portion of the observed $\delta^{18}O_{NO3+NO2}$ increase. Instead, the values are best explained by the presence of doubly-labeled ^{15}N-N_2O (i.e., $^{46}N_2O$) in the $^{15}N_2$ gas stocks. The apparent $\delta^{18}O_{NO3+NO2}$ of nitrate solutions equilibrated with all of the Sigma-Aldrich stocks, the Campro Scientific stock, and the *Cambridge B* $^{15}N_2$ stock proved to be greater than that of control solutions in the low sensitivity treatments (Fig. 1b). At 10 µmol L^{-1} nitrate, the apparent $\delta^{18}O_{NO3+NO2}$ of treated solutions was 188.5±83.8 among the *Sigma A1-A3* stocks (lot # SZ1670V), 169.8±17.9‰ for the *Sigma B* stock (lot # MBBB0968V), and 20.1±0.2 for the *Cambridge B* stock, compared to 18.9±0.3‰ in corresponding control solutions. The apparent $\delta^{18}O_{NO3+NO2}$ of the Campro Scientific stock at 10 µmol L^{-1} nitrate was 70.4±1.4‰, compared to 25.7±0.1‰ in corresponding control solutions. The apparent $\delta^{18}O_{NO3+NO2}$ of the samples decreased coherently with increasing nitrate concentrations for respective stocks. The apparent $\delta^{18}O_{NO3+NO2}$ values of solutions equilibrated with the *Cambridge A* stock, at 19.1±0.2‰, were not distinguishable from the control solutions. In the more sensitive equilibrations, nitrate solutions equilibrated with *Sigma B* $^{15}N_2$ gas had a $\delta^{18}O_{NO3+NO2}$ of 13,129±1186‰ compared to 23.9±0.2‰ in control solutions, whereas the $\delta^{18}O_{NO3+NO2}$

of the *Cambridge B* stock was 216.7±78.4‰ (Fig. 2b). Given that the apparent $\delta^{18}O_{NO3}$ enrichments are explained by the presence of $^{46}N_2O$, the inverse relationship between $\delta^{18}O$ values and nitrate concentration stems from the fact that the detected $^{46}N_2O$ derives from $^{46}N_2O$ dissolved in the nitrate solutions, and solutions containing higher nitrate concentrations require lower sample volume injections when using the denitrifier method for IRMS analysis. The observed excess $^{46}N_2O$ levels indicate $^{15}N^{15}N^{16}O$ contaminants (µmole $^{46}N_2O$ per mole of $^{15}N_2$) on the order of 41±21 among the *Sigma A1-A3* bottles, 49±17 in *Sigma B*, 11±3 in Campro Scientific, and 0.81±0.24 in *Cambridge B* (Table 1). The *Cambridge A* bottle was not tested in higher sensitivity dilutions that could have revealed the presence of some N_2O therein. The presence of $^{46}N_2O$ contaminant was verified directly for the *Sigma A2* lecture bottle from analyses of N_2O amended with injections of $^{15}N_2$ gas. Among four experimental samples, 109±5 µmoles of $^{46}N_2O$ were detected per mole of *Sigma A2* $^{15}N_2$ added, more than double the $^{46}N_2O$ that was detected in samples analyzed by the denitrifier method (39±8 µmole $^{46}N_2O$ per mole of $^{15}N_2$). This discrepancy likely resulted because samples analyzed by the denitrifier method were uncapped immediately prior to their injection into *P. aureofaciens* denitrifier cultures, allowing N_2O to escape to the atmosphere. As contaminant N_2O was not the target analyte of the denitrifier measurements, precautions were not taken to prevent N_2O gas loss at this step. The $^{46}N_2O$ concentrations derived from solution equilibrations of respective $^{15}N_2$ stocks thus constitute lower limits (Table 1).

Solutions equilibrated with Sigma-Aldrich $^{15}N_2$ gas showed substantial ^{15}N-enrichments of ammonium compared to control solutions (Fig. 3a): Equilibration with $^{15}N_2$ from the *Sigma A1* lecture bottle (lot # SZ1670V) resulted in a $\delta^{15}N_{NH4}$ of 99±39‰, compared to 0.6±0.5‰ for the control solution (NH_4Cl, SPEX CertiPrep); equilibration with $^{15}N_2$ from the *Sigma A2* bottle (lot # SZ1670V) yielded a $\delta^{15}N_{NH4}$ of 940±60‰, compared to 7.6±0.3‰ for the control solution (NH_4Cl salt, Fisher Scientific); equilibration with $^{15}N_2$ from the *Sigma B* bottle (lot # MBBB0968V) resulted in a $\delta^{15}N_{NH4}$ of 7030±2100‰, compared to 9.0±0.06‰ for the corresponding control solutions (NH_4Cl salt, Fisher Scientific). Mass balance calculations based on these isotope ratio values thus evidence the presence of 34±11, 518±26, and 1890±560 µmoles of ^{15}N-ammonium per mole of $^{15}N_2$ injected from *Sigma A1, A2*, and *B* $^{15}N_2$ bottles, respectively (Table 1). Unlike ^{15}N-labelled nitrate/

nitrite contaminants, the ^{15}N-ammonium contaminants appeared to be variable among bottles of lot # SZ1670V (Fig. 3a). The *Sigma A3* lecture bottle was not tested for ^{15}N-ammonium.

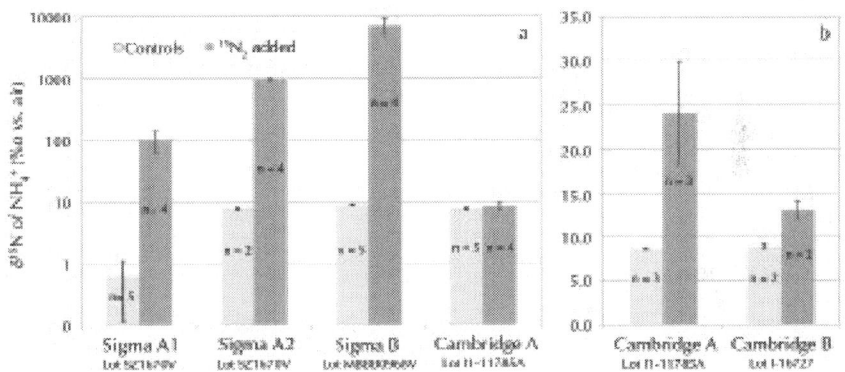

Figure 3: (a) $\delta^{15}N_{NH4}$ (log scale) of 5 μmol L^{-1} ammonium solutions after equilibration with 0.1 mL $^{15}N_2$ gas from respective Sigma-Aldrich and Cambridge Isotopes lecture bottles *vs.* control solutions. Sigma-Aldrich treatments utilized 40 mL ammonium solutions, whereas Cambridge Isotopes treatments utilized 100 mL ammonium solutions. (**b**) $\delta^{15}N_{NH4}$ of higher sensitivity equilibrations of 5 μmol L^{-1} ammonium solutions (10 mL) with 2.0 mL $^{15}N_2$gas from Cambridge Isotopes lecture bottles *vs.* control solutions. n = the number of experimental replicates.

In contrast to Sigma-Aldrich stocks, ammonium solutions equilibrated with $^{15}N_2$ from the*Cambridge A* bottle had a $\delta^{15}N_{NH4}$ of 8.3±1.0‰, comparable to that of the corresponding control solution of 7.6±0.3‰ (NH$_4$Cl salt, Fisher Scientific) in the lower sensitivity experiments (Fig. 3a). In the more sensitive dilutions, however, ^{15}N-ammonium contaminants were detected in both of the *Cambridge A* and *B* stocks (Fig. 3b). Solutions equilibrated with *Cambridge A* had a $\delta^{15}N_{NH4}$ of 24.0±5.9‰, compared to 8.7±0.1‰ for the control solutions (NH$_4$Cl salt, Fisher Scientific) and solutions equilibrated with *Cambridge B* had a $\delta^{15}N_{NH4}$ of 13.1±1.1‰, compared to 9.0±0.1‰ for the control solutions ((NH$_4$Cl salt, Fisher Scientific). The enrichment relative to control solutions invariably originates from a ^{15}N-ammonium contaminant, and cannot be attributed to a trace N_2O ($^{15}N^{14}N^{16}O$) contaminant, because the samples were purged when conducting ^{15}N-ammonium analyses, following the oxidation of ammonium to nitrite with hypobromite. These more sensitive treatments thus reveal the presence of minuscule ^{15}N-ammonium concentrations in the Cambridge Isotopes stocks, on the order 0.052±0.020 and 0.014±0.004 µmoles of ^{15}N-ammonium per mole of ^{15}N$_2$ gas in lots # I1–11785A and I-16727, respectively (Table 1).

The control solutions in the ^{15}N-ammonium experiments prepared from a single Fisher Scientific NH$_4$Cl salt stock revealed progressively heavier mean $\delta^{15}N_{NH4}$ values among experiments, at 7.6±0.3‰, 8.7±0.1‰, or 9.0±0.06‰. The solutions with a $\delta^{15}N_{NH4}$ estimated at 7.6±0.3‰ and 8.7±0.1‰ were made fresh from the salts for each experiment, such that the ^{15}N-enrichment of ammonium cannot be attributed to the progressive degassing of ammonia in solution during storage. In turn, isotopic standards for NH$_4^+$ (IAEA-N1 and N2) were stored in acidic solution, and thus were not subject to progressive degassing. Moreover, degassing of the isotopic standards would manifest as progressively lower δ^{15}N values measured for control solutions. Inter-batch variability intrinsic to the hypobromite-azide method [20] is plausible, as this technique is relatively recent, such that subtle sensitivities may not yet be apparent.

D. tertiolecta cultures grown in medium equilibrated with ^{15}N$_2$ gas from the *Sigma A3* bottle expectedly showed substantial ^{15}N enrichment of particulate nitrogen ($\delta^{15}N_{PN}$), averaging 44.3±6.1‰ among triplicate treatment cultures compared to 1.9±0.1‰ in control cultures (Fig. 4). Conversely, the $\delta^{15}N_{PN}$ of cultures equilibrated with *Cambridge B* ^{15}N$_2$ gas was 2.1±0.2‰, and thus not detectably different from that of control cultures, at 1.9±0.1‰ (Fig. 4).

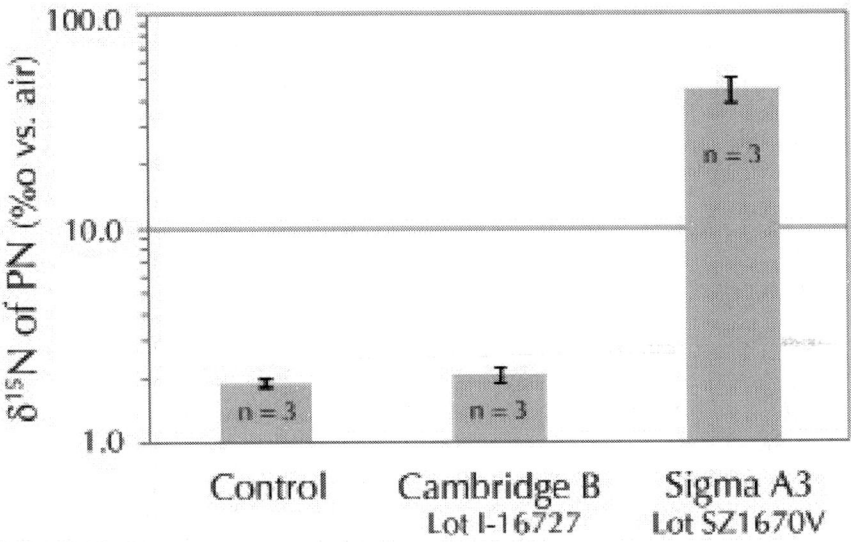

Figure 4: The $\delta^{15}N$ of particulate nitrogen ($\delta^{15}N_{PN}$) of *D. tertiolecta* harvested in stationary phase following growth in media containing sodium nitrate (and no ammonium) and equilibrated with $^{15}N_2$ gas from Sigma Aldrich or Cambridge Isotopes. n = the number of experimental replicates.

DISCUSSION

This study reveals that some commercial $^{15}N_2$ gas stocks contain contaminant ^{15}N-labeled bioavailable nitrogen species, including nitrate/nitrite, ammonium and nitrous oxide. Substantial levels of ^{15}N-labeled nitrate/nitrite, ammonium, and nitrous oxide were detected in Sigma-Aldrich stocks from lot # SZ1670V. Another Sigma-Aldrich stock from a different lot (# MBBB0968V) contained considerably less, but still significant, ^{15}N-nitrate/nitrite contaminants, similar nitrous oxide concentrations, and a greater concentration of ^{15}N-ammonium. Cambridge Isotopes stocks, in turn, contained relatively low concentrations of ^{15}N-nitrate/nitrite, ^{15}N-ammonium and ^{15}N-nitrous oxide. A $^{15}N_2$ stock from Campro Scientific contained no detected ^{15}N-nitrate/nitrite contaminants in low sensitivity experiments, but measurable ^{15}N-nitrous oxide. Indeed, a certificate of analysis provided by Campro Scientific attests that stocks may contain up to

15 ppm N_2O. ^{15}N-ammonium was not analyzed in Campro Scientific gas, nor was the stock tested at more sensitive $^{15}N_2$ dilutions, which could have revealed trace ^{15}N-nitrate or ammonium contaminants in the stock. In any case, ^{15}N-contamination with nitrous oxide is of no obvious consequence for biological $^{15}N_2$ applications, such as N_2 fixation rate measurements. However, the presence of ^{15}N-nitrate, nitrite and ammonium has serious implications for measurements of N_2 fixation, as these contaminants could lead to the detection of false positives or inflated rates.

The propensity of the detected ^{15}N-labeled contaminants to be assimilated into biomass was verified directly from cultures of *D. tertiolecta*, which acquired elevated $\delta^{15}N$ of PN in media equilibrated with a Sigma-Aldrich $^{15}N_2$ stock. Expectedly, media equilibrated with $^{15}N_2$ from a Cambridge Isotopes stock did not cause detectable ^{15}N-enrichment of biomass. At the given experimental conditions, however, the complete assimilation of the contaminant ^{15}N-nitrate/ nitrite from the Sigma-Aldrich stock should have yielded greater $\delta^{15}N_{PN}$ values than those observed, of at least $56.3 \pm 5.4‰$ (*vs.* control $\delta^{15}N_{PN}$ values of $\sim 1.9‰$), notwithstanding the additional contribution of any ^{15}N-ammonium contaminant (^{15}N-ammonium was not measured explicitly in the *Sigma A3* stock). This discrepancy is difficult to reconcile. We tentatively posit that ^{15}N-*nitrite* comprises a substantial fraction of the trace ^{15}N-nitrate/nitrite contaminant, and that *D. tertiolecta* may not be able to transport nitrite at nanomolar to sub-nanomolar concentrations. Indeed, such trace nitrite concentrations are likely below the thresholds achievable by micro-algal nitrite transport systems [24].

The contaminants in the $^{15}N_2$ stocks ostensibly derive from the method of $^{15}N_2$ gas production. $^{15}N_2$ gas is generally produced by the catalytic oxidation of ^{15}N ammonia ($^{15}NH_3$) gas with cupric oxide [25]. The bulk of the oxidation product is N_2 gas, although more oxidized N species are also produced in lesser quantities, specifically N_2O and NO [26]. Thus, potential contaminants in a $^{15}N_2$ gas stock would expectedly consist of unreacted ammonia gas, N_2O, and nitric oxide (NO). In contact with any oxygen and water vapor, NO would inadvertently be oxidized to nitric and nitrous acid [27], which would, in turn, dissociate to nitrate and nitrite upon dissolution in water, respectively. Purification of the $^{15}N_2$ gas from unreacted ammonia and from the generated nitrogen oxides involves sequential acid and alkaline scrubbing, respectively [28] and/or cryo-trapping of ammonia and NOx gases. Upon personal communication, Cambridge Isotopes and Campro Scientific did not provide details on their method of $^{15}N_2$ production, whereas Sigma-Aldrich reported that the company's subsidiary, Isotec, produces $^{15}N_2$ gas by the catalytic oxidation of ^{15}N-ammonia gas with cupric oxide, followed by sequential rounds of cryo-trapping and alkaline scrubbing of the N_2 gas to increase purity.

In order to gauge the extent to which the observed ^{15}N-ammonium contamination of Sigma-Aldrich and Cambridge Isotopes $^{15}N_2$ gas could skew estimates of N_2 fixation in incubations with $^{15}N_2$ gas, we modeled a field incubation experiment in which microorganisms assimilate the ^{15}N-ammonium contaminant rather than reduce $^{15}N_2$ gas. A simple finite-differencing model of a 'typical' oceanic N_2 fixation assay was devised, in which 0.1 mL of $^{15}N_2$ gas was equilibrated in two different water sample volumes, 0.25 L or 4.5 L, then incubated for 24 hours. Prescribed biomass and growth rates were characteristic of those at the oligotrophic surface ocean, namely, a particulate N stock of 0.2 µmol L^{-1} assimilating ammonium at a specific growth rate coefficient (µ) of 0.1 d^{-1} to 0.3 d^{-1}, with a recycling rate (the rate at which particulate N is returned to the ammonium pool) equivalent to the respective growth rate. The prescribed $\delta^{15}N$ of the initial particulate N was 0‰ [29], and the $\delta^{15}N$ of ambient ammonium was −2‰ [30]. Incremental initial concentrations of ambient ammonium were prescribed, from 1 nmol L^{-1} to 1 µmol L^{-1}. Ammonium concentrations in surface oligotrophic waters are typically very low (≤10 nmol L^{-1}), however, ammonium is a pervasive contaminant that could easily be introduced during sample preparation, as well as leached from incubation vial septa. We note that the ^{15}N-ammonium introduced by the $^{15}N_2$ gas, while substantial in terms of the $^{15}N/^{14}N$ ratio of ammonium, is on the order of ~20 nanomolar *at most* (under the modeled conditions), and thus has minimal effect on ambient ammonium concentrations. ^{14}N-ammonium contamination is expected to be negligible, given the method of $^{15}N_2$ gas synthesis. Finally, ^{15}N-ammonium assimilation was simulated for the broad range of ^{15}N-ammonium contaminant concentrations observed among Sigma-Aldrich and Cambridge Isotopes lecture bottles. N_2 fixation rates inferred from the simulated ^{15}N increase of particulate N were computed based on the formulation of Montoya (1996):

$$[PN]_\Delta/[PN]_f = (A_{PNf}A_{PN0})/(A_{N2}A_{PN0})$$

$$V = (1/\Delta t) * \left([PN]_\Delta/[PN]_f\right)$$

$$\rho = (V/2) * [PN]_f$$

$[PN]_\Delta$ is the change in particulate nitrogen concentration, $[PN]_f$ is the final particulate nitrogen concentration, A_{PNf} is the final ^{15}N enrichment of particulate nitrogen, A_{PN0} is the initial ^{15}N enrichment of particulate nitrogen, A_{N2} is the ^{15}N enrichment of the N_2 available for fixation, V is the specific rate of N_2 uptake, and ρ is the volumetric rate of N_2 fixation.

The model-derived 'N$_2$ fixation rates' resulting from Sigma-Aldrich ^{15}N-ammonium contaminant levels ranged from undetectable, <0.01 nmol N L^{-1} d^{-1}, to as high as 530 nmol N L^{-1} d^{-1} under the modeled conditions (Table 2). Rates were clearly sensitive to the concentration of ^{15}N-contaminant, the ambient ammonium concentration, the incubation volume, and the specific growth rate. At the lower level of ^{15}N-ammonium contaminant observed in the Sigma-Aldrich stocks, N_2 fixation rates were comparable to rates observed *in situ* for nearly all parameter permutations, from<0.01 to 9 nmoles N L^{-1} d^{-1}. N_2 fixation rates reported for marine environments cover a broad range, from 0.01 nmoles L^{-1} d^{-1} to tens of nmoles N L^{-1} d^{-1}[31]–[33]. Rates simulated with the highest observed level of contaminant, in smaller incubation volumes at given [15]N$_2$ additions, and/or with low ambient ammonium concentrations, tended to surpass rates observed *in situ* by 10 to 100 fold. The N_2 fixation rates modeled using the minute contaminant level detected in a Cambridge Isotopes stock ranged from undetectable to 0.02 nmoles N L^{-1} d^{-1} (Table 2), approximating the lower limit of some N_2 fixation rates reported in the literature [31],[34]–[38]. These simulated rates can be deemed conservative since the model does not account for any assimilation of contaminant ^{15}N-nitrate/nitrite, and the 0.1 mL [15]N$_2$ injection volume used in the model is on the lower end of [15]N$_2$ injection volumes typically used in open ocean N$_2$ fixation rate measurements.

Table 2: Inferred N_2 fixation rates (nmoles N $L^{-1} d^{-1}$) resulting from ^{15}N-labeled

Ambient (NH4+]	Cambridge Isotopes			Sigma-Aldrich			Sigma-Aldrich			
(μ mol L -1)	Cambridge A			Sigma A1			Sigma B			
	Lot # I1-11785A			Lot # SZI670V			Lot # MBBB0968V			
	052 gmol IsNH4-/mol 15N2			25 μm of I sNI-14-/mol IsN2			1900 mmol 15NH4 +/mol 15N2			
	μ =0.1	μ =0.2	μ =0.3	μ =0.1	μ =0.2	μ =0.3	μ =0.1	μ =02	μ =0.3	Incubation Volume (L)
0.001	0.019	0.019	0.019	9.0	9.0	9.0	310	460	530	0.25
0.01	0.014	0.016	0.016	73	83	8.6	250	400	470	
0.1	n.d.	n.d.	n.d	1.6	2.7	3.6	90	170	220	
1	n.d.	n.d.	n.d.	0.17	0.32	0.45	13	24	34	
0.001	n.d.	n.d.	n.d.	0.50	0.50	0.50	38	38	38	4.5
0.01	n.d.	n.d.	n.d.	0.42	0.47	0.48	30	35	36	
0.1	n.d.	n.d.	n.d.	0.08	0.15	0.19	6.5	11	15	
1	n.d.	n.d.	n.d.	n.d.	0.0 12	0.0 17	0.7	1.4	1.9	

N_2 fixation rates that would be inferred from 24-h field N_2 fixation assays conducted with $^{15}N_2$ stocks containing the respective concentrations ^{15}N-ammonium contaminants detected in Sigma-Aldrich and Cambridge Isotopes $^{15}N_2$ gas. In the simulations, microbial plankton assimilate ^{15}N-ammonium rather than fix $^{15}N_2$. Incubations are simulated in volumes of 0.25 L or 4.5 L equilibrated with 0.1 mL of $^{15}N_2$ gas, with 2.0 x10^{-7} pmol L^{-1} of plankton nitrogen (with a δ^{15}N = 0%o) assimilating at a range of specific growth rates, !I ($c1^{-1}$), countered by equivalent recycling rates, at incremental concentrations of ambient ammonium ($\delta^{15}N_{NH4}$= -2%o). Inferred rates of <0.01 nmoles N L^{-1} $c1^{-1}$ are considered undetectable (n.d.), doi:10.1371/journal.pone.0110335.t002

Based on the simulations above, the likelihood of N_2 fixation rates being inflated when using contaminated $^{15}N_2$ gas stocks is high It is surprising, then, that contamination of the $^{15}N_2$ stocks has not been reported previously. While growth solely upon N from N_2 fixation would eliminate the effect of ^{15}N-labeled bioavailable contaminants, it is expected that nitrate and ammonium assimilation would be rapid relative to N$_2$ fixation due to the prohibitive energetic cost of N$_2$ fixation [39]. A review of pertinent literature reveals that soil scientists were once aware of the possible contamination of ^{15}N$_2$ with bioavailable N, and took steps to mitigate it [28,40]. However, to the best of our knowledge, there is no mention of potential contamination of ^{15}N$_2$ stocks in the marine literature, or in more recent terrestrial literature. The fact that this issue has gone unnoticed could mean that major contamination of ^{15}N$_2$ gas stocks, such as that observed here in Sigma-Aldrich stocks, could be limited to the current lots. Supporting the notion that contamination is rare is the observation of undetectable N$_2$ fixation rates at the surface ocean, where phytoplankton readily assimilate ammonium [38] – even in investigations utilizing the Sigma-Aldrich (Isotec) ^{15}N$_2$ gas [31]. However, a representative at Isotec stated that their procedures for synthesis and purification of ^{15}N$_2$ gas have not changed in past decades, which suggests that ^{15}N-contaminants may have been pervasive in previous lots. Failure to detect interferences from ^{15}N-contaminants in previous studies may then stem from incubation conditions conspiring to yield expected rates of *apparent* N$_2$ fixation in spite of the presence of ^{15}N contaminants (Table 2). Interference of ^{15}N contaminants on N$_2$ fixation rate measurements may then be relatively minor in systems where bioavailable N assimilation rates are low and/ or where ambient nitrate and ammonium concentrations are relatively elevated ($\geq$100 nmol L^{-1}; Table 2), as ambient assimilable N effectively diminishes ^{15}N enrichment resulting from ^{15}N-labeled contaminants.

It is difficult, if not impossible, to discern whether N$_2$ fixation rate estimates in previous studies may have been confounded due to the assimilation of ^{15}N contaminants in ^{15}N$_2$ gas stocks. Given that ^{15}N$_2$ stocks from only one of the three suppliers tested here contained contaminants to an extent that would interfere with any but the lowest reported N$_2$ fixation measurements, there is a strong likelihood that published estimates performed with ^{15}N$_2$ from the other two suppliers have not been significantly inflated by labeled contaminants. In fact, many estimates may be lower than reality due to the incomplete equilibration of ^{15}N$_2$ gas with the incubation medium, a pervasive problem with aqueous ^{15}N$_2$ fixation assays that was diagnosed only recently [41]–[43]. Nevertheless, it is advisable at this point to analyze commercial ^{15}N$_2$ stocks prior to their use to ensure their relative purity.

In doing so, particular attention must be paid to the lower limit of detection for N_2 fixation rates. In recent years, workers have reported estimates of very low rates (≤ 0.1 nmol L^{-1} d^{-1}) in environments where N_2 fixation is otherwise unexpected, which include oxygen-deplete regions of the water column at Pacific margins[34,35,38], as well as in the Beaufort Gyre of the Arctic Ocean [36]. Such minimal rates are questionable, considering that the relatively clean Cambridge Isotopes $^{15}N_2$ gas was found to contain enough ^{15}N-ammonium to infer N_2 fixation rates of up to 0.02 nmoles N L^{-1} d^{-1}. Campro Scientific and other commercially available $^{15}N_2$ gas stocks could similarly contain minute, but significant, concentrations of ^{15}N-nitrate or ammonium. Therefore, it behooves investigators to not only verify the purity of their commercial $^{15}N_2$ prior to use, but also to generate constraints as to the lower limit of detection, allowing for the possibility that a trace-level ^{15}N-contaminant could interfere with the detection of diminutive N_2 fixation rates.

Steps toward Mitigation

The catalytic synthesis of $^{15}N_2$ gas from ^{15}N-ammonia gas invariably entails the incidence of ^{15}N-ammonium and ^{15}N-NOx contaminants, the removal of which is dependent on the stringency of scrubbing procedures to which a given batch is subjected. The consistency of ^{15}N-nitrate/nitrite measurements among bottles from an individual lot from Sigma-Aldrich (*Sigma A1-A3*), in contrast to the lower ^{15}N-nitrate/nitrite detected in a subsequent lot (*Sigma B*), supports the premise that the levels of ^{15}N-contaminants are associated with discrete batch syntheses, identified by lot numbers, rather than with individual lecture bottles. The variability in ^{15}N-ammonium among lecture bottles of the same lot suggests that ammonia gas does not disperse homogeneously in compressed N_2 gas. In any case, large-scale batch syntheses of $^{15}N_2$ occur relatively infrequently, on the order of every 2 years at Isotec (subsidiary of Sigma-Aldrich). We currently have a verbal agreement with Isotec to perform nitrate and ammonium isotopic analyses of $^{15}N_2$ batches, toward providing a certificate of analysis ensuring adequate purity for N_2 fixation assays. In the meantime, we advise that workers procure low-contaminant stocks from lots that we tested here. The very high purity of the batches from these suppliers suggests stringent and efficacious purification protocols, such that batches synthesized by these groups in the future are likely to be equally pure, notwithstanding the potential for human error during synthesis or subsequent purification.

Regardless of 'expected' purity, we recommend that workers explicitly test new batches availed by respective suppliers for ^{15}N-nitrate and ammonium prior to using them in N_2 fixation assays, and actively disseminate the results to targeted web-based forums. To test a given batch, ^{15}N$_2$ gas can be equilibrated with nitrate and ammonium solutions following protocols akin to the low and high sensitivity equilibrations herein. A number of laboratories perform commercial nitrate isotope analyses routinely at a modest cost per sample. Ammonium isotope analyses are substantially more involved, but are also performed routinely by a number of laboratories.

We further recommend that pertinent publications include not only the brand of ^{15}N$_2$ stock, but also the associated lot number, and references to reported contaminants. With continued testing, our understanding of the prevalence of commercial ^{15}N$_2$ contamination will grow and shed light on this problem, which may have plagued N_2 fixation estimates in the past.

ACKNOWLEDGMENTS

We wish to thank Nicole Chang for assistance with ^{15}N$_2$ measurements, David Cady at the University of Connecticut for assistance with particulate nitrogen and carbon δ^{15}N and δ^{14}C measurements, and Mark Rollog and Thomas Kuhn for assistance in the lab at the University of Basel.

AUTHOR CONTRIBUTIONS

Conceived and designed the experiments: RD JG CRT MAA MFL PHM. Performed the experiments: RD JG LT MFL Analyzed the data: RD JG CRT LT MFL Contributed reagents/materials/analysis tools: JG CRT MFL Contributed to the writing of the manuscript: RD JG MFL PHM LT CRT MAA

REFERENCES

1. Montoya JP, Voss M, Kahler P, Capone DG (1996) A simple, high-precision, high-sensitivity tracer assay for N_2 fixation. Appl Environ Microbiol 62(3): 986–93.

2. Stewart WDP, Fitzgerald GP, Burris RH (1967) In situ studies on N_2 fixation using the acetylene reduction technique. Proc Natl Acad Sci USA 58(5): 2071–8. doi: 10.1073/pnas.58.5.2071

3. Hardy RWF, Holsten RD, Jackson EK, Burns RC (1968) The acetylene-ethylene assay for N_2 fixation: laboratory and field evaluation. Plant Physiol 43(8): 1185–207. doi: 10.1104/pp.43.8.1185

4. Burris RH, Miller CE (1941) Application of N^{15} to the study of biological nitrogen fixation. Science 93(2405): 114–5. doi: 10.1126/science.93.2405.114

5. Hardy RWF, Burns RC, Holsten RD (1973) Applications of the acetylene-ethylene assay for measurement of nitrogen fixation. Soil Biol Biochem 5(1): 47–81. doi: 10.1016/0038-0717(73)90093-x

6. Giller KE (1987) Use and abuse of the acetylene-reduction assay for measurement of associative nitrogen fixation. Soil Biology & Biochemistry 19(6): 783–4. doi: 10.1016/0038-0717(87)90066-6

7. Staal M, Lintel-Hekkert ST, Harren F, Stal L (2001) Nitrogenase activity in cyanobacteria measured by the acetylene reduction assay: A comparison between batch incubation and on-line monitoring. Environ Microbiol 3(5): 343–51. doi: 10.1046/j.1462-2920.2001.00201.x

8. Belay N, Sparling R, Choi B, Roberts M, Roberts J, et al. (1988) Physiological and ^{15}N-Nmr analysis of molecular nitrogen fixation by methanococcus-thermolithotrophicus, methanobacterium-bryantii and methanospirillum-hungatei. Biochim Biophys Acta 971(3): 233–45. doi: 10.1016/0167-4889(88)90138-3

9. Scharff A, Egsgaard H, Hansen P, Rosendahl L (2003) Exploring symbiotic nitrogen fixation and assimilation in pea root nodules by in vivo ^{15}N nuclear magnetic resonance spectroscopy and liquid chromatography-mass spectrometry. Plant Physiol 131(1): 367–78. doi: 10.1104/pp.015156

10.   Addison SL, McDonald IR, Lloyd-Jones G (2010) Identifying diazotrophs by incorporation of nitrogen from $^{15}N_2$ into RNA. Appl Microbiol Biotechnol 87(6): 2313–22. doi: 10.1007/s00253-010-2731-z

11.   Guillard RRL, Ryther JH (1962) Studies of marine planktonic diatoms: I. Cyclotella nana Hustedt, and Detonula confervacea (cleve) Gran. Can. J. Microbiol. 8: 229–239. doi: 10.1139/m62-029

12.   Braman R, Hendrix S (1989) Nanogram nitrite and nitrate determination in environmental and biological-materials by vanadium (III) reduction with chemi-luminescence detection. Anal Chem 61(24): 2715–8. doi: 10.1021/ac00199a007

13.   Holmes R, Aminot A, Kerouel R, Hooker B, Peterson B (1999) A simple and precise method for measuring ammonium in marine and freshwater ecosystems. Can J Fish Aquat Sci 56(10): 1801–8. doi: 10.1139/cjfas-56-10-1801

14.   Sigman D, Casciotti K, Andreani M, Barford C, Galanter M, et al. (2001) A bacterial method for the nitrogen isotopic analysis of nitrate in seawater and freshwater. Anal Chem 73(17): 4145–53. doi: 10.1021/ac010088e

15.   Casciotti K, Sigman D, Hastings M, Böhlke J, Hilkert A (2002) Measurement of the oxygen isotopic composition of nitrate in seawater and freshwater using the denitrifier method. Anal Chem 74(19): 4905–12. doi: 10.1021/ac020113w

16.   Gonfiantini R (1984) Report on an advisory group meeting on stable isotope reference samples for geochemical and hydrochemical investigations. Vienna, 19–21 Sept. 1983. IAEA, Vienna.

17.   Böhlke JK, Coplen TB (1995) Interlaboratory comparison of reference materials for nitrogen isotope ratio measurements. Proceedings of a consultants meeting held in Vienna, 1–3. Dec. 1993, IAEA-TECDOC-825, 51–66.

18.   Böhlke J, Mroczkowski S, Coplen T (2003) Oxygen isotopes in nitrate: new reference materials for ^{18}O:^{17}O:^{16}O measurements and observations on nitrate-water equilibration. Rapid Commun Mass Spectrom 17(16): 1835–46. doi: 10.1002/rcm.1123

19. McIlvin MR, Casciotti KL (2011) Technical updates to the bacterial method for nitrate isotopic analyses. Anal Chem 83(5): 1850–6. doi: 10.1021/ac1028984

20. Zhang L, Altabet MA, Wu T, Hadas O (2007) Sensitive measurement of $NH_4{}^{+15}N/{}^{14}N$ ($^{15}NH_4{}^+$) at natural abundance levels in fresh and saltwaters. Anal Chem 79(14): 5297–303. doi: 10.1021/ac070106d

21. Böhlke J, Gwinn C, Coplen T (1993) New reference materials for nitrogen-isotope-ratio measurements. Geostand Newsl 17(1): 159–64. doi: 10.1111/j.1751-908x.1993.tb00131.x

22. Kendall C, Grim E (1990) Combustion tube method for measurement of nitrogen isotope ratios using calcium-oxide for total removal of carbon-dioxide and water. Anal Chem 62(5): 526–9. doi: 10.1021/ac00204a019

23. Qi H, Coplen T, Geilmann H, Brand W, Bohlke J (2003) Two new organic reference materials for ^{13}C and ^{15}N measurements and a new value for the ^{13}C of NBS 22 oil. Rapid Communications in Mass Spectrometry 17(22): 2483–7. doi: 10.1002/rcm.1219

24. Cordoba F, Cardenas J, Fernandez E (1986) Kinetic characterization of nitrite uptake and reduction by chlamydomonas-reinhardtii. Plant Physiol 82(4): 904–8. doi: 10.1104/pp.82.4.904

25. Bergersen FJ (1980) Measurements of nitrogen fixation by direct means. In: Bergersen FJ, editor. Methods for evaluating biological nitrogen fixation. Wiley-Interscience, Chichester, pp. 5–110.

26. Il'chenko N (1976) Catalytic-oxidation of ammonia. Usp Khim 45(12): 2168–95. doi: 10.1070/rc1976v045n12abeh002765

27. Ashmore PG, Burnett MG, Tyler BJ (1962) Reaction of nitric oxide and oxygen. Trans Faraday Soc 58: 685–691. doi: 10.1039/tf9625800685

28. Ohyama T, Kumazawa K (1981) A simple method for the preparation, purification and storage of $^{15}N_2$ gas for biological nitrogen fixation studies. Soil Sci Plant Nutr 27(2): 263–5. doi: 10.1080/00380768.1981.10431278

29. Altabet M (1988) Variations in nitrogen isotopic composition between sinking and suspended particles - implications for nitrogen cycling and particle transformation in the open ocean. Deep-Sea Research Part A-Oceanographic Research Papers 35(4): 535–54. doi: 10.1016/0198-0149(88)90130-6

30. Fawcett SE, Lomas M, Casey JR, Ward BB, Sigman DM (2011) Assimilation of upwelled nitrate by small eukaryotes in the Sargasso Sea. Nat Geosci 4(10): 717–22. doi: 10.1038/ngeo1265

31. Dore J, Brum J, Tupas L, Karl D (2002) Seasonal and interannual variability in sources of nitrogen supporting export in the oligotrophic subtropical North Pacific Ocean. Limnol Oceanogr 47(6): 1595–607. doi: 10.4319/lo.2002.47.6.1595

32. Montoya J, Holl C, Zehr J, Hansen A, Villareal T, et al. (2004) High rates of N_2 fixation by unicellular diazotrophs in the oligotrophic Pacific Ocean. Nature 430(7003): 1027–31. doi: 10.1038/nature02824

33. Needoba JA, Foster RA, Sakamoto C, Zehr JP, Johnson KS (2007) Nitrogen fixation by unicellular diazotrophic cyanobacteria in the temperate oligotrophic North Pacific Ocean. Limnol Oceanogr 52(4): 1317–27. doi: 10.4319/lo.2007.52.4.1317

34. Fernandez C, Farias L, Ulloa O (2011) Nitrogen fixation in denitrified marine waters. PLoS One 6(6): e20539. doi: 10.1371/journal.pone.0020539

35. Hamersley MR, Turk KA, Leinweber A, Gruber N, Zehr JP, et al. (2011) Nitrogen fixation within the water column associated with two hypoxic basins in the Southern California Bight. Aquat Microb Ecol 63(2): 193–205. doi: 10.3354/ame01494

36. Blais M, Tremblay J, Jungblut AD, Gagnon J, Martin J, et al. (2012) Nitrogen fixation and identification of potential diazotrophs in the Canadian Arctic. Global Biogeochem Cycles 26: GB3022. doi: 10.1029/2011gb004096

37. Halm H, Lam P, Ferdelman TG, Lavik G, Dittmar T, et al. (2012) Heterotrophic organisms dominate nitrogen fixation in the South Pacific Gyre. Isme Journal 6(6): 1238–49. doi: 10.1038/ismej.2011.182

38. Dekaezemacker J, Bonnet S, Grosso O, Moutin T, Bressac M, et al. (2013) Evidence of active dinitrogen fixation in surface waters of the eastern tropical South Pacific during El Nino and La Nina events and evaluation of its potential nutrient controls. Global Biogeochem Cycles 27(3): 768–79. doi: 10.1002/gbc.20063

39. Stam H, Stouthamer A, Vanverseveld H (1987) Hydrogen metabolism and energy costs of nitrogen-fixation. FEMS Microbiol Lett 46(1): 73–92. doi: 10.1111/j.1574-6968.1987.tb02453.x

40. De-Polli H, Matsui E, Döbereine J, Salati E (1977) Confirmation of nitrogen fixation in two tropical grasses by $^{15}N_2$ incorporation. Soil Biol and Biochem 9: 119–123. doi: 10.1016/0038-0717(77)90047-5

41. Mohr W, Grosskopf T, Wallace DWR, LaRoche J (2010) Methodological underestimation of oceanic nitrogen fixation rates. PLoS One 5(9): e12583. doi: 10.1371/journal.pone.0012583

42. Grosskopf T, Mohr W, Baustian T, Schunck H, Gill D, et al. (2012) Doubling of marine dinitrogen-fixation rates based on direct measurements. Nature 488(7411): 361–4. doi: 10.1038/nature11338

43. Wilson ST, Boettjer D, Church MJ, Karl DM (2012) Comparative assessment of nitrogen fixation methodologies, conducted in the oligotrophic North Pacific Ocean. Appl Environ Microbiol 78(18): 6516–23. doi: 10.1128/aem.01146-12.

Chapter 3

Potential of Best Practice to Reduce Impacts from Oil and Gas Projects in the Amazon

Matt Finer[1], Clinton N. Jenkins[2], and Bill Powers[3]

[1]Biodiversity Program, Center for International Environmental Law, Washington D.C., United States of America

[2]Department of Biology, North Carolina State University, Raleigh, North Carolina, United States of America

[3]E-Tech International, Santa Fe, New Mexico, United States of America

ABSTRACT

The western Amazon continues to be an active and controversial zone of hydrocarbon exploration and production. We argue for the urgent need to implement best practices to reduce the negative environmental and social impacts associated with the sector. Here, we present a three-part study aimed at resolving the major obstacles impeding the advancement of best practice in the region. Our focus is on Loreto, Peru, one of the largest and most dynamic hydrocarbon zones in the Amazon. First, we develop a set of specific best practice guidelines to address the lack of clarity surrounding the issue. These guidelines incorporate both engineering-based criteria and key ecological and social factors. Second, we provide a detailed analysis of existing and planned hydrocarbon activities and infrastructure, overcoming the lack of information that typically hampers large-scale impact analysis. Third, we evaluate the planned activities and infrastructure with respect to the best practice guidelines. We show that Loreto is an extremely active hydrocarbon front, highlighted by a number of recent oil and gas discoveries and a sustained government push for increased exploration. Our analyses reveal that the use of technical best practice could minimize future impacts by greatly reducing the amount of required infrastructure such as drilling platforms and access roads. We also document a critical need to consider more fully the ecological and social factors, as the vast majority of planned infrastructure overlaps sensitive areas such as protected areas, indigenous territories, and key ecosystems and watersheds. Lastly, our cost analysis indicates that following best practice does not impose substantially greater costs than conventional practice, and may in fact reduce overall costs. Barriers to the widespread implementation of best practice in the Amazon clearly exist, but our findings show that there can be great benefits to its implementation.

INTRODUCTION

The western Amazon, one of the most biologically and culturally rich regions on Earth [1]–[3], continues to be an active and controversial zone of hydrocarbon exploration and production [4]. Hydrocarbon blocks – geographic areas delimited by national governments for the exploration and production of oil and gas – cover vast swaths of the region, including protected areas and titled indigenous territories [5]. Moreover, international bidding rounds on new oil and gas blocks in Colombia, Ecuador, and Peru confirm that exploration activities continue expanding deeper into the most remote tracts of the western

Amazon. The lone exception is Ecuador's Yasuní-ITT Initiative, a novel government proposal that seeks international compensation in exchange for not drilling sizable oil deposits in the core of the megadiverse Yasuní National Park [1], [6].

With governments promoting ever more oil development in the western Amazon, there needs to be greater attention given to minimizing the associated ecological and social risks [7]. Direct impacts include deforestation for access roads, drilling platforms, helipads, and pipeline routes, as well as contamination from spills, leaks and discharges [5]. Indirect effects, which include selective logging, hunting, and deforestation, primarily arise from the human colonization along new access routes [5]. Considerable social conflict, particularly with native communities, may also arise from these direct and indirect impacts [5].

While we strongly support efforts like the Yasuní-ITT Initiative as a potential mechanism to avoid completely the problems of hydrocarbon activities in the Amazon, we also argue for rigorous best practices where projects do move forward. We define a best practice as one that minimizes the environmental impact associated with typical practice, and that has been successfully employed in a commercial oilfield exploration or production project in Latin America.

At least three major obstacles currently impede the advancement of best practice in the western Amazon. First, best practice lacks a precise set of guidelines in applicable regulations. This regulatory gray area allows project proponents to define almost any practice as "best practice," and often results in typical high-impact practice being approved as best practice in environmental impact studies. Second, the lack of easily accessible and precise data on planned activities and infrastructure makes it difficult for policy makers and civil society to evaluate upcoming projects and push for best practice. Much of the currently available information relates to just the geographic extent of the hydrocarbon blocks, and not the more important planned activities within. Third, questions regarding cost, or assumptions that best practice will impose substantially greater costs, are common and likely deter companies from deviating from conventional practices.

We present here a three-part study aimed at overcoming these obstacles and demonstrating the potential of hydrocarbon sector best practice to minimize ecological and social impacts in western Amazonia. Our focus is on the Department of Loreto in northern Peru

(Figure 1). Loreto, along with the neighboring Ecuadorian Amazon, is one of the largest and most dynamic hydrocarbon zones in the Amazon [5], [8].

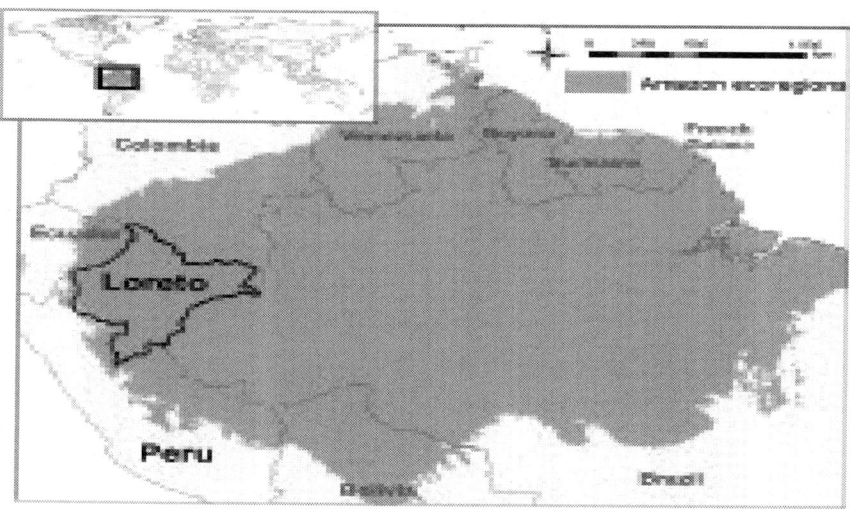

Figure 1: Study focal area. We focus on the Department of Loreto in the northern Peruvian Amazon. Amazon ecoregions are as defined by [61].

Loreto, a vast territory covering nearly 369,000 km², makes an ideal case study for a number of reasons. The region possesses extraordinary biological and cultural diversity [1], [9], along with vast tracts of largely intact tropical forest, driving an urgency to minimize extractive industry impacts. It is home to a large number of active hydrocarbon blocks spanning the full range of project stages, from pre-exploration to long-time production. In regards to the latter, a pair of 1970s-era oil operations caused significant contamination by dumping toxic production waters into local waterways for nearly four decades [10]. Therefore, local policy makers and residents are acutely aware of the potential risks from oil development. In addition, a number of recent exploration projects have yielded new oil and gas discoveries in Loreto, greatly increasing the probability that hydrocarbon development will continue as a major issue for the region well into the future.

We first present a set of best practice guidelines designed to minimize the impact of hydrocarbon activity in the Amazon. These guidelines incorporate both engineering-based criteria and key ecological and social factors. E-Tech International originally formulated the engineering guidelines, which are based on both Peruvian law and the latest in global technology [11]. We subsequently added the ecological and social factors to ensure that engineering best practice projects also do not threaten sensitive areas.

Second, we provide a detailed analysis of existing and planned hydrocarbon activities and infrastructure. In doing so, we move beyond evaluation based solely on the extent of hydrocarbon blocks and provide a more comprehensive examination of actual activities. This includes detailed data on existing and planned activities for all field-based phases of a hydrocarbon project, namely seismic exploration, exploratory wells, production wells, access roads, and pipelines.

Third, we evaluate the planned activities and infrastructure with respect to the best practice guidelines from part one. We analyze all planned projects in relation to both the engineering guidelines and the following four ecological and social factors: protected areas, indigenous territories, critical ecosystems, and priority watersheds. This evaluation represents a more strategic, larger-scale analysis than the current system of project-level, local-scale studies, and it would ideally take place within the context of a Strategic Environmental Assessment (SEA)[5]. Since 2008, Peruvian law has required national, regional, and local authorities to undertake SEAs for plans, polices, and programs that may have significant environmental impacts [12],[13], but only a handful have been completed to date [14].

We also conduct an initial analysis on the estimated difference in cost between use of best practice and conventional development.

Finally, we discuss our findings in terms of how the use of best practice can minimize negative impacts, particularly deforestation and contamination.

RESULTS

Best Practice

The basis of the best practice guidelines was an analysis of both cutting-edge technology and Peruvian regulation (Table 1). To understand the implementation of best practice, it is important to understand first the typical life cycle of a hydrocarbon project in the Peruvian Amazon, which follows several basic steps. The government agency Perupetro creates the blocks ("lotes" in Spanish) and then promotes and auctions them internationally [15]. Recently there have been annual or biannual bidding rounds with one to two dozen blocks promoted and auctioned together. Perupetro ultimately signs the final contract with the selected company for each respective block, but the contract must first be approved by presidential decree [15]. The contract term, which runs 30 years for oil and 40 years for natural gas, includes two phases: exploration and production. The exploration phase is for seven years (with possible extensions) and includes a Minimum Work Program for the required amount of seismic lines and exploratory wells to be carried out by the operating company [15].

Table 1: Best practice guidelines

1.Presentation of an overall project development plan based on best practice prior to initiating the exploration phase.
2.Use of state-of-the-art subsurface computer model that integrates airborne electromagnetic data and existing seismic data to minimize the need for new seismic projects.
3.All exploration and production platforms must be capable of drilling Extended Reach Drilling (ERD) wells with a horizontal displacement of at least 8 km (i.e., minimum distance between platforms of 16 km).
4.New access road construction is prohibited (e.g., no new roads between platforms and processing facilities or in pipeline/flowline rights-of-way).
5.Permanent camps may only be constructed along the banks of navigable rivers, not in the jungle interior.
6.Only permissible means of transport are by air and river, with defined limits on the size of transport vessels and on frequency of movements.
7.The maximum pipeline/flowline right-of-way construction width must be less than 13 m with intervals of canopy bridges at least every 1,000 m.

8.Pipelines should be designed/operated with: increased wall thickness to withstand soil movements and effects of internal erosion; regular internal traverses with intelligent inspection tools to detect internal abnormalities and lateral movement of the pipeline; automatic shut-off valves at each tie-in point of welded pipeline sections; and oil spill rapid response teams.
9.Adequate funds must be reserved for site abandonment that includes removal and/or remediation of contaminated materials, soil, and water sources, and revegetation of cleared areas with native species.
10.Consideration of key ecological and social factors such as protected areas, indigenous territories, key ecosystems, and key watersheds in determining whether oil & gas development should be pursued at all.

Two types of seismic testing are common in the Amazon, 2-dimensional (2D) and 3-dimensional (3D) [5], [11]. The former generates an initial 2D cross-section of the subsurface, while the latter generates a 3D model to define in detail the deposit(s). On the ground, 2D is characterized by relatively spread-out linear transects (at least 1 km separation) cut through the forest, whereas 3D lines form tight grids (100s of meters separation) and are typically measured in square kilometers [11]. Seismic lines are typically less than two meters wide and do not require the cutting of large trees. Explosive charges are placed at regular intervals along these lines in holes of six to nine meters, and parallel lines of geophones register the echo patterns of the explosions on subsurface structures. These echo patterns reveal geologic structures that may contain oil or gas and that may warrant further assessment with exploratory wells [11].

If commercially viable quantities of oil or gas are discovered, the concession may proceed to production phase. However, contracts may be, and often are, terminated by the operating company during the exploration phase. Historically, the design of production phase has been characterized by many closely spaced drilling platforms, extensive networks of access roads, and pipeline routes with wide right-of-ways [11]. Moreover, in a number of projects designed during the 1970s, traditional practice included the dumping of toxic production waters directly into local waterways.

Engineering Criteria

The first step of best practice, from an engineering perspective, is that the operating company must present an overall conceptual plan based on best practice for all phases of the project before beginning any work on the ground. We recommend that such a best practice conceptual plan be required during the company submission of its Minimum Work Program to the government during the bidding phase. This system would have the dual benefit of incorporating best practice into the bidding competition and subsequently the final contract signed by the company and the government. As a result, the use of best practice would be a formal and binding obligation. This recommendation of incorporating best practice into the Minimum Work Program would require a modification to current regulation.

Following this step, exploration activities should combine remote aerial electromagnetic surveys of subsurface structures with existing field information to create a precise state-of-the-art subsurface computer model of the hydrocarbon structures. The construction of this model involves an integrated approach that uses existing field data from seismic testing and exploratory wells as calibration points for new remote sensing data. A recent project in Brazil demonstrated the utility of this integrated approach to produce a precise subsurface computer model with minimal new intervention on the ground [11], [16]. The aim of this innovation is to conduct new seismic testing only in areas where there is a demonstrated potential for commercial deposits. Typically oil companies do not combine the remote sensing data with existing data from earlier exploration programs to refine the study area for the purpose of minimizing the amount of subsequent seismic testing.

At the core of best practice is Extended Reach Drilling (ERD), a technique to reach a larger subsurface area from one surface drilling location. First developed in the late 1980s, ERD is a type of advanced directional drilling where the horizontal reach is at least two times greater than the vertical depth [11]. In practical terms, it means a single drilling platform can reach multiple distant targets in an oil or gas deposit, thereby reducing the total number of required platforms. The U.S. National Petroleum Council [17] recently recognized ERD as a key technology for reducing footprints of drilling operations. The current world record for ERD is 12.4 km, and any horizontal distance up to 8 km is now considered routine for an ERD well [11]. Therefore, there should be a large separation, at least 16 km, between drill sites.

ERD has been used in numerous Latin American exploratory and production drilling projects, but not yet in the Peruvian Amazon. In Argentina, two recent exploration projects employed ERD wells with horizontal displacements of approximately 4 and 5 km, in 2007 and 2008 respectively [11]. Also in Argentina, a production project beginning in 1997 drilled a series of ERD wells of more than 10 km. Most recently, in 2011, an exploration project in Colombia employed an ERD well. Although ERD has not yet seen application in Peru, it is important to note that national hydrocarbon regulation does require that drilling sites disturb the least amount of land possible [18] (see Article 67). Use of ERD would minimize the amount of land disturbed for drilling sites compared to any typical project limited to vertical or directional drilling techniques only.

The use of ERD relates to two additional key best practices: 1) no new access roads, processing facilities, or permanent camps beyond the banks of navigable rivers, and 2) transport of people, materials, and equipment must be by air or river (with controls on size and frequency of movements). In other words, companies must operate as if at sea, a roadless development concept known as the offshore model [19]. In addition, production platforms deeper in the jungle and away from navigable rivers must be unmanned, with raw production fluids transported via roadless flowlines to the respective processing facility located along a navigable river. Processing facilities are where the production fluids – oil, gas, and production water – are separated, and the oil is prepared for export via pipeline, the gas burned for onsite use, and the production water re-injected into a subsurface formation. These points related to roadless development are consistent with Peruvian hydrocarbon regulation, which requires preferential use of river and air transport, and which states that road construction can only proceed if it is demonstrated that river and air transport are not possible [18] (see Article 40). For example, the Camisea natural gas project in southern Peru has been in operation since 2004 with no permanent camps away from navigable rivers and no access roads [11].

Regarding pipelines and flowlines, best practice calls for a greatly minimized right-of-way (ROW), with a reduction from the traditional 25 m down to 13 m or less. This "green pipeline" ROW technique, or "ducto verde" in Spanish, also emphasizes conforming the ROW to natural contours and emphasis on manual clearing (instead of heavy machinery) to further reduce impacts, particularly on steep slopes. This type of reduced-impact pipeline corridor was employed on one ROW section of the Camisea Project, in contrast to the higher-impact traditional pipeline ROWs used in other pipeline/flowline sections of the same project. Another major advantage of this type of narrowed ROW corridor is the ability to maintain canopy bridges. Canopy bridges are tree canopy sections along the ROW that remain intact to facilitate the passage of wildlife, at intervals of approximately one kilometer or more [20]. In order to minimize contamination threats related from pipelines, best practice also calls for increased wall thickness (to withstand soil movements and internal erosion), regular internal traverses with intelligent inspection gauges to detect internal abnormalities and lateral movement of the pipeline, automatic shut-off valves at each welded tie-in point, and establishment of rapid response teams [21].

In terms of site abandonment, companies must set aside adequate funds to assure removal and/or remediation of contaminated materials, soil, and water sources, and revegetation of cleared areas with native species [11].

Ecological and Social Factors

In addition to the engineering-based best practices, it is critical to consider a range of key ecological and social factors. In other words, using technical best practice is not necessarily a license to operate in sensitive areas. Based on previous evaluations of ecological and social factors to consider in assessing projects in areas of high biodiversity and intact forest [5], [8], [22], [23], we chose five: protected areas, priority watersheds, key ecosystems, indigenous territories, and proposed reserves for indigenous peoples in voluntary isolation.

Loreto has 14 official protected areas as established by the national protected areas agency SERNANP. Of these, 11 are managed nationally (two national parks, four national reserves, two communal reserves, and three reserved zones) and 3 are managed regionally (regional conservation areas). In the IUCN system of protected area categories, Peruvian national parks are considered as category II, national reserves as category VI, and the remaining areas either have no category or it is currently undeclared. Of these five types of protected area designations, just national parks are off-limits to extractive industries according to Peruvian Law. However, the new Güeppi – Sekime National Park (established in October 2012) allows the continuation of previously existing concessions. Therefore, 13 of the 14 protected areas in Loreto do not legally prohibit hydrocarbon activities. However, the national protected areas agency (SERNANP) must provide a technical favorable opinion before the energy ministry will approve activities within protected areas.

For priority watersheds, we focus on the Nanay River, a critical resource that provides drinking water to the departmental capital city of Iquitos. The classification of additional priority watersheds in Loreto is still under review by authorities. For key ecosystems, we focus on white-sand forests. Although low in overall species diversity, this rare and fragile ecosystem contains a high number of endemics and is considered a high conservation priority in Loreto[24].

Loreto is also home to a great abundance of indigenous peoples' territories. According to the latest publicly available data from the Instituto del Bien Común (IBC), there are around 500 titled indigenous territories in Loreto. Data for solicited new territories or solicited extensions of existing territories are more preliminary. The IBC data indicate that there are 24 solicited new territories and 29 solicited extensions of existing territories, although the true figures are likely to be much higher for both. In addition, within Loreto there are five proposed reserves for indigenous peoples in voluntary isolation. The right of indigenous peoples to be consulted in order to obtain their free, prior and informed consent about development decisions that will affect them is established under the International Labor Organization's Convention 169 [25] and the United Nations Declaration on the Rights of Indigenous Peoples [26]. Peru is a signatory of the former and voted in support of the latter. Moreover, Peru promulgated a landmark indigenous consultation law based on ILO 169 in 2011 [27].

Finally, two additional factors to consider, but beyond the scope of this study, are the greenhouse gas emissions and use of royalties from hydrocarbon activities. Regarding the former, carbon emissions arise from project-related forest loss, transportation, and energy generation, and of course the ultimate burning of the extracted hydrocarbons [28].

Indeed, one of the selling points of Ecuador's Yasuní-ITT Initiative is not only the avoided on-site deforestation, but also the maintenance of 410 million metric tons of CO_2 permanently underground [6]. For the latter, it is important to note that over 90% of royalties from hydrocarbon activities go to regional and local governments, and a portion of this money is used for transportation and other development projects that may also have environmental and social impacts [29], [30].

Existing and Planned Activities and Infrastructure

Hydrocarbon Blocks

As of October 2012, there were 48 hydrocarbon blocks in Loreto (Figure 2), covering 215,169 km² or 57.4% of the department. Of these, 29 are active concessions under contract with multinational energy companies. Four of these active concessions are in production phase (Blocks 1AB, 8, 31B, and 67) and the remaining 25 in exploration phase. The remaining 19 blocks are part of Perupetro's new bidding round.

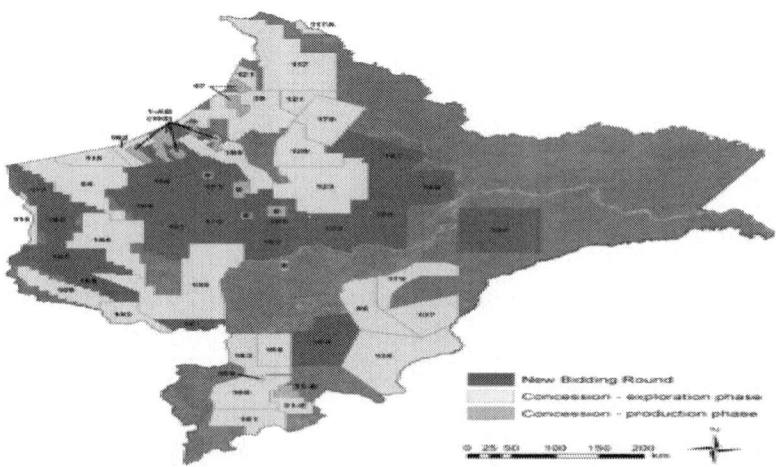

Figure 2: Hydrocarbon blocks in Loreto. There are three general types of blocks based on the contractual agreement between government and a com-

pany: concession in exploration phase, concession in production phase, and proposed concession under promotion or negotiation.

Of the 25 concessions in the exploration phase, five have approved or pending environmental impact studies for seismic testing, three for exploratory wells, and six for both seismic testing and exploratory wells. The remaining concessions have not yet prepared environmental impact studies or begun exploration work.

Twenty-nine companies were operating or participating in the Loreto concessions during 2012. All but one are multi-nationals based outside of Peru. The 28 multi-nationals originate from 14 countries, including Argentina, Brazil, Canada, Colombia, France, Spain, Vietnam, the United Kingdom, and the United States of America. However, company turnover is relatively high. For example, during the course of this study, the primary concession holder changed in Blocks 64, 67, 123, and 129.

There are two important additional items to emphasize regarding this current state of hydrocarbon blocks in Loreto. First, although at the time of this publication Block 67 was not yet producing oil, the operating company declared this block commercially viable in late 2006, and it is currently officially classified as production phase. Second, many hydrocarbon blocks have previously existed but subsequently been retired and do not appear in Figure 2. Thus, many exploration wells and seismic lines displayed in subsequent figures appear outside the current blocks.

Seismic Testing.

Oil companies have conducted extensive 2D seismic testing in Loreto over the past 40 years, with a smaller but increasing amount of 3D seismic testing in recent years (Figure 3). This includes 61,403 km of 2D seismic lines (9% conducted since 2007) and 2,565 km^2 of 3D seismic (71% conducted since 2007). As illustrated in Figure 3, testing has been concentrated in southern and central Loreto, while much of northern and eastern Loreto has yet to experience major exploration. In regards to planned testing, five blocks (95, 109, 121, 130, and 135) have pending 2D projects totaling 3,900 km (Figure 3). Two additional blocks (1AB and 39) have pending 3D projects totaling 1,738 km^2.

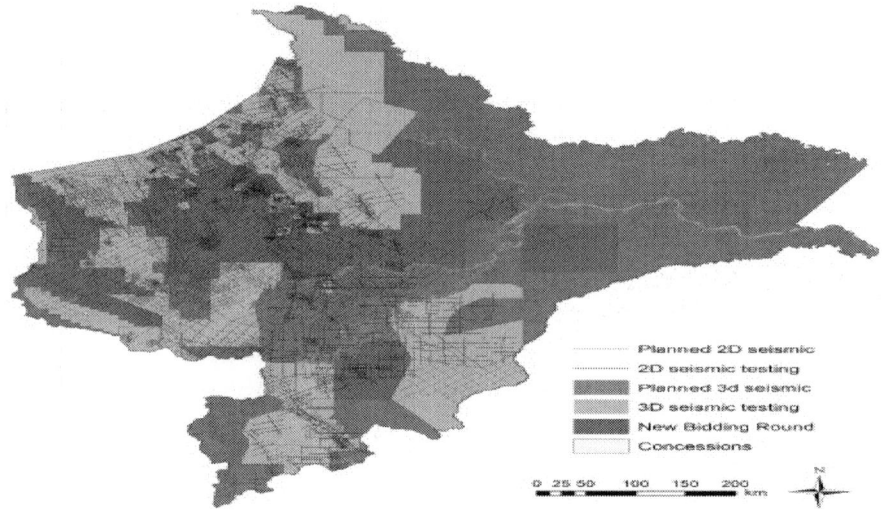

Figure 3: Existing and planned 2D and 3D seismic testing in Loreto. 2D testing is represented by straight lines and is measured in kilometers while 3D testing is represented by polygons and measured in square kilometers. It is important to note that numerous hydrocarbon blocks have previously existed but subsequently been retired. Thus, many seismic lines appear outside the current blocks.

Exploratory and Production Wells

Official data indicate that oil companies have drilled 223 exploratory wells in Loreto (Figure 4A), with 12% of them drilled since 1998 (the earliest date for which we have detailed data). Of these wells, nearly half (105) are outside of current production blocks and therefore may provide key field information to create subsurface computer models, potentially minimizing the need for extensive new exploratory campaigns.

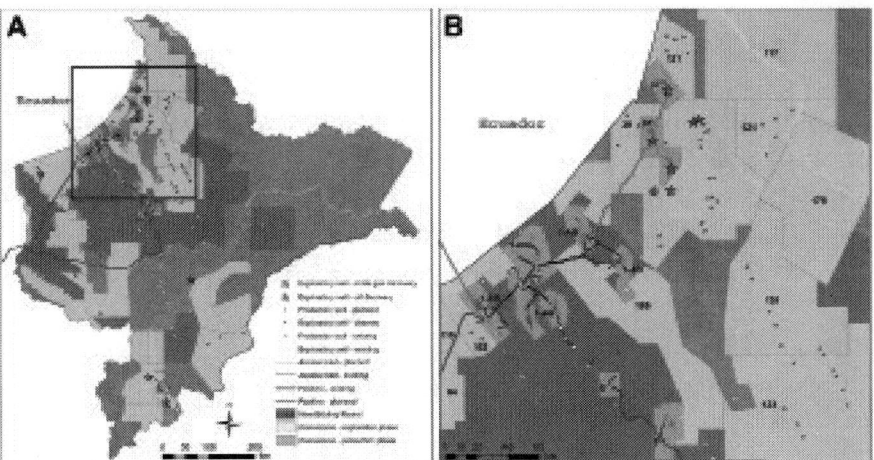

Figure 4: Existing and planned exploratory wells, production wells, access roads, and flowlines/pipelines in Loreto. (A) Map for all of Loreto. Note that stars indicate the Block 64 light crude oil discovery, the Block 95 medium oil discovery, and the Block 31 shale gas discovery. (B) Zoom of high activity zone in and around Blocks 1AB, 39, and 67. Note that stars indicate the Blocks 39 and 67 heavy crude oil discoveries. It is important to note that numerous hydrocarbon blocks have previously existed but subsequently been retired. Thus, many exploration wells appear outside the current blocks.

Companies operating in Loreto have extracted 1.016 billion barrels of oil [31]. Of this production, Blocks 1AB, 8, and 31 have contributed 68%, 31%, and 1%, respectively. Annual oil production in Loreto peaked at 47 million barrels in 1979 [32] and has steadily fallen to 10.2 million barrels in 2011 [31], [33], [34], a decrease of 78%. The Peruvian Energy Ministry estimates over 393 million barrels of oil remain in these blocks (72% in Block 1AB, 25% in Block 8, and 3% in Block 31) [35].

There are currently 219 active production wells in Loreto (Figure 4A). Most are in Block 1AB (62.5%) (Figure 4B), with the remainder in Block 8 (20.5%) and Block 31 (17%). According to the Energy Ministry, there are also ~50 active reinjection wells and ~240 inactive and abandoned production wells [31].

Seventeen of the 28 exploratory wells drilled since 1998 have encountered hydrocarbon deposits in Blocks 31E, 39, 64, 67, 95, and 100 (Figure 4A). The type of newly discovered hydrocarbon varies considerably, with light oil in Block 64, medium oil in Block 95, heavy oil in Blocks 39 and 67, and shale gas in Block 31E. The Peruvian Energy Ministry estimates reserves (proven, probable, and possible) of around 928 million barrels in the three blocks with oil (40.5% in Block 39, 35% in Block 67, and 24.5% in Block 64) [35]. An additional 31.6 million barrels is reported from the latest oil discovery in Block 95. In terms of upcoming production, Block 67 is by far the most advanced, with approved environmental impact studies for the pipeline and development wells (Figure 4B).

Environmental impact studies have been submitted for 66 additional exploratory well platforms in Blocks 39, 64, 95, 102, 121, 123, 129, 130, and 135 (Figure 4A).

Roads and Pipelines.

Over time, the companies operating Blocks 1AB and 8 have constructed an extensive access road and flowline network to service the production wells and processing facilities. In addition, the North Peruvian Pipeline transports oil from these blocks to Peru's Pacific coast. Within Loreto, this flowline/pipeline network extends ~1,156 km (Figure 4A). Transport of crude oil from Block 31 to Pucallpa is via the Ucayali River.

There are plans to extend the existing pipeline network to connect with the new oil discoveries in the region (Figure 4A). The environmental impact study for a new 207 km pipeline to transport heavy crude from Block 67 to the starting point of the existing North Peruvian Pipeline was approved in 2011 (Figure 4B). Completion of this pipeline is scheduled for 2017. Preliminary plans also exist to transport light crude from Block 64 to the existing North Peruvian Pipeline. In August 2012, Peru and Ecuador signed an agreement that would allow the transport of Ecuadorean crude across the border to the North Peruvian Pipeline.

We calculate a cumulative network of 803 km of access roads in Blocks 1AB, 8, and 31 (Figure 4A). The largest access road network by far is in Block 1AB, a sprawling network of 504 km (Figure 4B).

The recently approved environmental impact study for the Block 67 production wells includes plans for a new 85 km access road network adjacent to the internal pipelines during the construction phase (Figure 4B). According to the approved plan, half of this access road network will be eliminated after the construction phase, including the connections between the three oil fields. There are also preliminary plans for construction of a 36 km access road in Block 64 (Figure 4A).

Best Practice

Engineering Criteria.

We analyzed all planned projects in relation to the best practice guidelines presented earlier. Starting with Block 67 as an example, the production plan consists of 21 production well platforms and three processing facilities (Figure 5a). The platforms are distributed among the three major oil deposits (eight in Paiche, six in Dorado, and seven in Piraña) and each major deposit has its own processing facility. Within each oil deposit, the multiple production platforms are located relatively close together, often separated by less than two km. In each case, the proposed drilling platforms are all within eight km of a single hypothetical ERD-capable drilling platform (Figure 5a). Figure 5b illustrates an alternative ERD-based Block 67 production field design using just three ERD-capable drilling platforms, one for each oil deposit. In addition, note that in Figure 5b, the Dorado processing facility is gone, leaving only the two processing facilities, Paiche and Piraña, located near navigable rivers. This important modification would also mean the elimination of nearly all access roads.

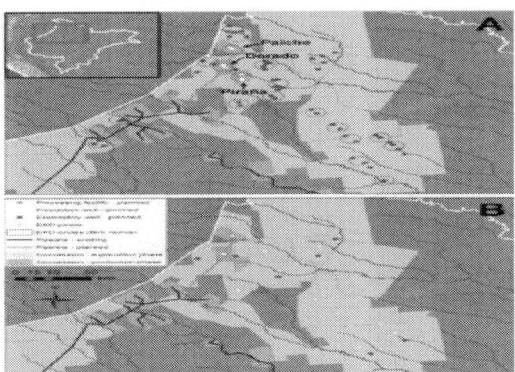

Figure 5: Analysis of planned projects in relation to best practice guidelines in high activity zone of Loreto. (A) Planned exploratory wells, production

wells, and processing facilities. Red circles indicate multiple platforms within eight kilometers of a single hypothetical ERD-capable drilling platform. (B) Alternative design based on best practice. Yellow dots indicate where an ERD-capable drilling platform could replace multiple planned platforms. Note that the Dorado processing facility is eliminated due to distance from a navigable river.

In regards to the planned Block 67 pipeline, the main technical issue is the width of the right-of-way. As noted above, best practice calls for a maximum ROW width of 13 m or less. The Block 67 operating company at the time (Perenco) originally proposed a 25 m ROW width for much of the pipeline length (177 km), with a reduction to 20 m for the 30 km length crossing the Pucacuro National Reserve. Under pressure from the Peruvian government, the company increased the section having a 20 m width to 141 km, but resisted committing to the 13 m width achievable with the green pipeline technique. The company also agreed to a reduced ROW width of 10 m for 0.68 km of the pipeline corridor within Pucacuro that will include canopy bridges.

We also analyzed all other current plans for new exploratory well drilling platforms to determine how many proposed platforms could be eliminated by employing ERD. Figure 5b illustrates an alternative scenario that eliminates all platforms within 16 km of each other, with exploratory wells being drilled up to 8 km from each platform. This scenario assumes each drilling platform is ERD-capable. Of the 66 planned platforms in Blocks 39, 64, 95, 102, 121, 123, 129, 130, and 135, we estimate that nearly half (31) could be eliminated using ERD.

This reduction in infrastructure would translate directly to a reduction in deforestation. According to a sampling of environmental impact studies, we found that each new drilling platform requires the clearing of 2 to 4.5 hectares of forest and production phase processing stations require around 6 hectares each. For example, the Block 67 development project without best practice – consisting of 3 processing stations and 21 drilling platforms – would require a footprint exceeding 1 km^2 for these facilities. Using best practice to eliminate 18 drilling platforms and one processing facility would reduce forest loss by over 75%. In addition, the new Block 67 access road network and pipeline corridor, without best practice, would result in an additional 7 km^2 of direct forest loss [36], [37]. With best practice, total direct forest loss would be significantly less, as the vast majority of the roads would be eliminated and the pipeline corridor would be seven to twelve meters narrower along nearly the entire length.

In addition, a review of environmental impact studies and post-project reports reveals that best practice would result in reduced forest loss during the exploration seismic phase. Most seismic projects require at least 50 heliports (larger projects may call for at least 200) and literally hundreds of camps and drop zones [8]. Typical area requirements are around 2,400 m² for helipads, 300 m² for temporary camps, and 20 m² for drop zones. For example, a recently completed 1,480 km 2D seismic operation in Blocks 123 and 129 (that constructed 272 heliports, 208 camps, and 4,050 drop zones) had a cumulative 0.85 km² footprint [38], [39]. A planned 3,700 km 3D seismic operation in Block 39 (calling for 75 heliports, 42 camps, and 3,800 drop zones) projects a 5.99 km² footprint [40].

Ecological and Social Factors

We found that oil blocks overlap 34% (29,000 km²) of the protected area system in Loreto, with 19 blocks overlapping 10 protected areas (eight national and two regional). The protected areas that are the most compromised by oil blocks include the Alto Nanay – Pintuyacu – Chambira Regional Conservation Area, Sierra del Divisor Reserved Zone, and Pucacuro National Reserve. A number of blocks cover an additional 17,150 km² of officially designated protected area buffer zones, primarily around Cordillera Azul National Park and Pacaya-Samiria National Reserve. A number of currently producing wells in Block 8 are within Pacaya-Samiria. Two of the recent Block 39 oil discoveries are within Pucacuro, as is a 30 km stretch of the planned pipeline from Block 67 to the Northern Peru Pipeline. There are 21 planned exploration wells within three protected areas. Thirteen of these planned wells are within Alto Nanay – Pintuyacu – Chambira (Blocks 123 and 129), seven are within Pucacuro (Block 39), and one is within Sierra del Divisor (Block 135).

The vast majority of blocks (90%) overlap titled or petitioned indigenous territories. Put another way, the oil blocks overlap 68% of these indigenous lands (42,548 km²). Production wells in Blocks 1AB, 8, and 31 are located around or upstream of indigenous communities. This is also true of the oil discovery in Block 64 and seven additional planned exploratory wells in other blocks.

Twelve oil blocks overlap 60% (21,962 km²) of the proposed reserves for indigenous peoples in voluntary isolation. Note that there is an extremely high level of existing and planned activity within the proposed Napo-Tigre Territorial Reserve. The three recently discovered Block 67 oil deposits, two of the Block 39 oil discoveries, and 48 planned exploratory wells are within the reserve. There are also three planned exploration wells in the proposed Yavari-Tapiche Territorial Reserve.

Twelve blocks overlap white-sand forest patches. Indeed, blocks cover all of the known large patches of white-sand forest outside of Allpahuayo-Mishana National Reserve. Several wells in Blocks 123, 129, and 135 are close to white-sand forests. The Northern Peru Pipeline crosses one of the largest white-sand forest patches.

Finally, fourteen planned exploratory wells in Blocks 123 and 129 are within the Nanay watershed, as are sections of four new blocks included in the new bidding round.

When combining all areas covered by protected areas, indigenous territories, white-sand forests, and the Nanay watershed, we found that nearly half (48%) of the total hydrocarbon block area in Loreto overlaps at least one key ecological or social factor (Figure 6). In addition, 80% of the planned exploratory wells, 100% of the planned production platforms, most of the recent hydrocarbon discoveries, and 59% of the planned pipelines contain such an overlap.

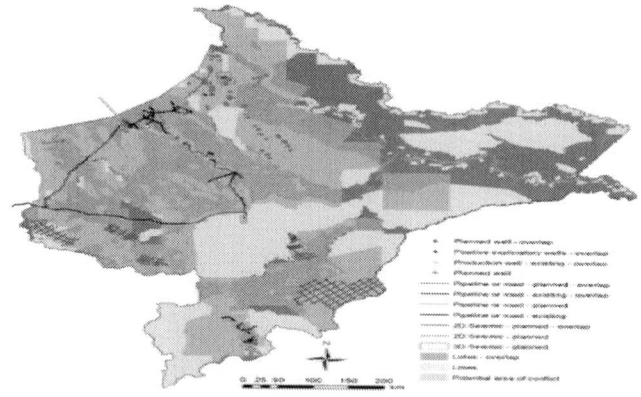

Figure 6: Consideration of key ecological and social factors: overlaps. Light blue indicates an important or sensitive area that is not covered by a hydro-

carbon block, while orange indicates an area that is covered by a block. Further, we indicate planned 2D and 3D seismic testing, exploratory and production wells, access roads, and flow lines/pipelines that would overlap with at least one of the key ecological and social factors.

Cost Analysis.

There is enough data available on costs for the planned Block 67 development project to make a comparison between the proposed conventional project and an alternative project using best practice [11], [41]. This cost analysis considered changes due to use of ERD, elimination of the access road network, elimination of one processing facility, and implementation of the green pipeline ROW construction technique.

The average depth of the wells in Block 67 is approximately two kilometers. Therefore, only a well with a horizontal displacement of greater than four kilometers would be considered an ERD well. Assuming a single, central drilling platform in each of the three oil fields, we estimated that one-third of the planned wells would use ERD and the remaining two-thirds would be conventional directional wells. A conventional well costs $3.5 million and the cost of an ERD well increases approximately linearly with its horizontal displacement. Therefore, assuming that an ERD well's average horizontal displacement will be twice that of a conventional well, we estimated an average ERD well cost of $7 million. We calculated that the use of ERD for one-third of the wells would increase costs by about $220 million.

Several other key components of best practice, however, would reduce costs. The elimination of 18 planned drilling platforms due to use of ERD would reduce costs by about $142 million. The elimination of one of the planned processing facilities would reduce costs by about $36.5 million (this estimate includes additional costs for expanding one of the other planned processing facilities to accept more flow).

In terms of transportation costs, the elimination of the access road network would reduce costs by about $45 million. Reliance on extensive jungle road networks and diesel-fueled heavy vehicles, using imported diesel fuel, adds a substantial operational cost. There would be some increase in helicopter flights, though this expense would be offset by the near-elimination of heavy vehicle traffic in the block. In regards to arriving to the site, barges already move on regular schedules from Iquitos to docks of active concessions throughout Loreto and therefore do not represent a major new expense.

Overall, we found that best practice does not translate to substantially higher costs, and may in fact reduce total expenses. The operating company for Block 67 estimated total costs of $1.339 billion [41]. We estimate that total costs for the best practice alternative is $1.321 billion.

DISCUSSION

Loreto, a vast region larger than Germany or nearly the size of Montana, is one of the most active and dynamic hydrocarbon zones in the Amazon. Forty-eight oil blocks cover over half the department, an affected area of over 215,000 km². These blocks cover the full range of project development stages: 4 in production, 25 in various stages of exploration, and the remaining 19 are part of Perupetro's latest international bidding round. Adding to the complexity, 29 companies operate the production and exploration phase blocks, and company turnover is frequent.

Companies have extracted over one billion barrels of oil from Loreto over the past 40 years. However, a major long-term trend of decreasing production has spurred efforts to boost exploration in search of additional deposits. This trend will begin to reverse with the imminent start of production in Block 67, the most recent block to enter into production phase. Two additional recent notable discoveries include heavy oil in Block 39 and light oil in Block 64. The Peruvian Energy Ministry estimates reserves of over 900 million barrels of oil in these three blocks. Together with the remaining reserves in Blocks 1AB and 8, Loreto may have another billion barrels of oil available.

A key wild card is the shale gas discovery in Block 31E. This discovery is significant because of the potentially large size of the shale formation, the novelty of developing this type of gas deposit in Peru, and the possible utilization of shale fracturing techniques [42]. Recent experience in the United States has demonstrated that there are significant and unique risks associated with shale gas production, and that these risks are not yet fully understood [43].

More new discoveries are likely given that exploration activities remain very active. Indeed, 44 of the 48 blocks in Loreto are in either exploration or bidding phase, 13 of which already have finalized environmental studies for seismic testing and exploratory wells. In other words, extensive and widespread amounts of exploration are still to come.

Impacts and the Role of Best Practice

With such a large number of hydrocarbon projects, it is critical to advance best practice as a means of minimizing social and environmental impacts in Loreto. The original design and operations of Blocks 1AB and 8 – characterized by many closely-spaced drilling platforms, dumping toxic production waters directly into local waterways, and extensive access road networks – represent high-impact, 1970s-era technology [11]. In contrast, best practice incorporates a number of technological advances and strategic planning techniques to minimize negative impacts, such as deforestation and contamination.

We demonstrated that the use of technical best practice, in the case of Block 67, would reduce impacts by: 1) reducing the number of drilling platforms from twenty-one to three, 2) eliminating one of the three processing facilities, 3) eliminating virtually the entire access road network, and 4) narrowing the pipeline right-of-way. Furthermore, we estimate that the use of ERD-capable drilling rigs across all exploration blocks in Loreto could eliminate about half of the proposed drilling platforms. In the context of a Strategic Environmental Assessment, this would represent a lower-impact, "greener" scenario, in relation to the higher-impact Business-As-Usual scenario.

We further found that this reduction in infrastructure from best practice would directly translate to a reduction in deforestation. In the case of Block 67, forest loss would drop by around 50, 75, and 100% from drilling platforms and processing facilities, the pipeline, and access roads, respectively. Moreover, the reduction of access roads could prevent substantial secondary deforestation. Fortunately, the isolated existing access roads have not yet triggered significant indirect forest loss from subsequent colonization and logging, as roads have in the neighboring Ecuadorian Amazon. If connected to the rest of Peru's road network, as called for in long-term government plans, indirect deforestation would likely quickly escalate.

The reduction in drilling platforms by employing best practice may also serve to reduce contamination. Blocks 1AB and 8 resulted in nearly four decades of significant contamination through the dumping of toxic production waters into local waterways, until indigenous inhabitants forced an accelerated phase-out of this practice between 2006 and 2009 [10], [44]. However, pollution problems continue to plague local communities, as all three current oil producing blocks in Loreto (Blocks 1AB, 8, and 31B) have had major leaks and spills in recent years [45], [46]. In addition to the now mandatory practice of reinjecting toxic production waters, best practice serves to reduce contamination by significantly reducing the number of point sources (i.e., drilling platforms) and designing more strategic flowline/pipeline routes.

Our best practice guidelines also aim to minimize the negative impacts from exploration phase seismic testing. Our review of environmental impact studies and post-project reports revealed that traditional seismic projects do cause deforestation, primarily from the need to construct hundreds of helipads, temporary camps, and drop zones. In addition, seismic testing, particularly the more intensive 3D form, results in helicopter noise, an inux of workers, the cutting of hundreds of kilometers of seismic lines through the understory, and the detonation of thousands of underground seismic charges [47]. A recent study found a significant decrease in the group sizes of the endangered white-bellied spider monkey (*Ateles belzebuth*) during 2D seismic testing in Block 39 [48], although these same researchers found no negative impacts on ocelots (*Leopardus pardalis*) [49].

As part of best practice, we contend that the extent of future seismic testing, and therefore its associated impacts, could be greatly reduced by combining existing exploration data with remote sensing data in a state-of-the-art subsurface computer model. The region has already been subject to over 61,000 km of 2D seismic testing, 2,500 km² of 3D seismic testing, and 220 exploratory wells. However, companies operating in the region typically do not analyze this existing information in combination with remote sensing data for the purpose of minimizing the amount of new seismic testing. Instead, extensive new seismic testing programs are still the norm, as evidenced by the more than 3,400 km of planned 2D seismic and 1,700 km² 3D seismic projects. Given the extensive amount of existing exploration data in Loreto, this modeling advance offers a methodology that may greatly minimize the extent of new seismic campaigns.

We also raised the important need to consider ecological and social factors in addition to technical best practice criteria. We found that nearly half of the total block area and the vast majority of planned exploration wells, production platforms, and planned pipeline length overlap sensitive areas in Loreto. For example, oil blocks overlap over one-third of the protected area system, two-thirds of the titled and solicited indigenous territories, nearly all of the large white-sand forest patches, and nearly the entire Nanay watershed. Recognizing and minimizing these types of conflictive overlaps early in the government's concession evaluation process could avoid future conflicts. For example, the current controversy over planned exploratory wells in the Nanay watershed, the source of the capital city's water supply, could have been avoided by excluding this area from concessions in the first place. However, history may be doomed to repeat itself as four of the new bidding round blocks overlap this same watershed.

Identifying overlaps and possible conflicts with indigenous communities is also an important element of the new indigenous consultation law. This law, which entered into force in April 2012, is debuting in Loreto with the re-leasing of Block 1AB as Block 192 (current contract expires in 2015). Indigenous organizations are demanding a number of important actions, such as the remediation of existing environmental damages, resolution of land-titling disputes, and consultation with affected indigenous communities before the bidding process begins [50]. They are also calling for the elaboration of a Strategic Environmental Assessment for all planned and existing blocks.

Finally, we demonstrated that incorporating best practice does not impose substantially greater costs than a conventional project, and may in fact reduce overall costs. Although costs for ERD wells are around double that of conventional wells, the reduction in costs from elimination of drilling platforms, access roads, and remote processing facilities counterbalance the higher well construction costs.

Large barriers to the widespread implementation of best practice in Loreto and the rest of the Amazon clearly exist. Despite meetings and letters urging Peruvian officials to mandate use of ERD and green pipeline ROW in Block 67, the environmental impact studies were approved without full adoption of these key elements of best practice. Further work is needed to advance the concepts discussed in this paper, ideally in the form of a government-led Strategic Environmental Assessment.

METHODS

We obtained all GIS data described below from existing sources, no field work was conducted in this study. However, in some cases we revised the data if obvious differences were observed in satellite imagery.

Analysis of Existing and Planned Activities and Infrastructure

We obtained GIS data for hydrocarbon blocks, seismic lines, exploratory wells, and pipelines from Perupetro in November 2011 and October 2012. We acquired GIS data for production wells from Perupetro in July 2012. Additional information on seismic testing, exploratory wells, production wells, oil production, and operating companies is from monthly "Informe Estadístico" and yearly "Anuario Estadístico" reports available on the Ministerio de Energía y Minas website (http://www.minem.gob.pe). We acquired information on whether or not recent exploratory wells encountered hydrocarbon deposits from a Perupetro presentation [51] and press reports. We updated the status of the blocks using the environmental impact studies published on the Ministerio de Energía y Minas website. Data pertaining to the new bidding round blocks are from information included in a Perupetro presentation [52].

For existing pipelines, additional GIS data are from the Loreto Regional Government. We compared the Petroperu pipeline datasets to recent Landsat and higher resolution satellite imagery in Google Earth and ArcGIS basemaps to produce a revised pipeline layer. This revised layer included route corrections for known pipelines and the addition of spurs visible in the satellite imagery but not included in either of the original datasets.

For existing access roads, we obtained two GIS datasets. The first was from the national government via the Ministerio de Transportes y Communicaciones. The second was from the Loreto Regional Government. We compared both datasets to recent Landsat and higher resolution satellite imagery in Google Earth and ArcGIS basemaps to produce a revised data layer. This revised layer included route corrections and the addition of spurs visible in the satellite imagery but not included in either of the original datasets.

Data for planned seismic lines and exploratory wells are from environmental impact studies published on the Ministerio de Energía y Minas website. Information related to the planned production wells in Block 67 is from the relevant environmental impact study [37]. For planned pipelines, we obtained information from the relevant Block 67 environmental impact studies[36], [37], a public presentation by a Block 64 operating company representative in Iquitos, Peru (June 2012), and press reports regarding the pipeline extension to Ecuador. For planned access roads, information is from the relevant Block 67 environmental impact study [37] and an operating company report detailing development options for Block 64.

The cut-off date for incorporating new data was March 2013.

Best Practice

We analyzed all planned projects in relation to both the engineering guidelines and identified ecological and social factors. For the engineering criteria component, we identified all planned exploratory wells and production platforms that are within eight kilometers of a single central drilling platform. These wells could therefore be drilled from a central drilling platform using an ERD-capable drilling rig. We also identified all river sections with at least 5,000 upstream cells in Hydro SHEDS [53], which we used as a proxy for year-round navigability of the river. This data was used to corroborate the feasibility of limiting permanent camps and processing facilities to sites along navigable rivers. For the estimates on avoided deforestation, we collected information on the area required for drilling platforms, processing facilities, and seismic activities from a sampling of current environmental impact studies and post-project reports from Blocks 39, 67, 102, 123, 127, 128, 129, 130 and 135.

For the ecological and social factors component, we analyzed all existing and planned activities and infrastructure in relation to:

protected areas, indigenous territories, white-sand forest patches, and the Nanay watershed. Data for protected areas are from SERNANP [54]. Subsequently we digitized three new areas created after the data were obtained from SERNANP. GIS data for indigenous territories are from the Instituto del Bien Común [55]. Data for white sand forest patches are from Nature Serve, Field Museum, and published studies [24],[56], [57]. Analyses were done in ArcGIS 10.1.

For the comparative cost analysis, we used oil industry guidelines on the definition of ERD wells of at least 2:1 ratio of horizontal displacement to vertical depth [58] and the relative cost of an ERD well (proportionate to length of well), industry data on the maximum length of oil and natural gas flowlines [59], and a comparative cost estimate of green pipeline and conventional pipeline ROW construction costs [60]. Specific data for the Block 67 case study came from the actual projected costs estimated by the operating company (Perenco) to fully develop Block 67. These costs were presented in an official environmental impact study response by Perenco [41]and approved by the Energy Ministry in January 2012. This document includes details on the cost of all major Block 67 infrastructure elements, including well development, drilling platforms, processing facilities, permanent camps, roads, docks, and logistical bases.

ACKNOWLEDGMENTS

We thank Carroll Muffett, Melissa Blue Sky, Amanda Kistler, Enrique Ortiz, and three anonymous reviewers for reviewing the manuscript and providing helpful feedback. We thank Valeria Urbina, Patricia Patron, Asunta Santillan, and Cristina López for valuable comments and assistance in gathering data.

AUTHOR CONTRIBUTIONS

Conceived and designed the experiments: MF CNJ BP. Performed the experiments: MF CNJ BP. Analyzed the data: MF CNJ BP. Contributed reagents/materials/analysis tools: MF CNJ BP. Wrote the paper: MF CNJ BP.

REFERENCES

1. Bass MS, Finer M, Jenkins CN, Kreft H, Cisneros-Heredia DF, et al.. (2010) Global Conservation Significance of Ecuador's Yasuní National Park. PLoS ONE 5(1): e8767. Available: http://www. plosone.org/article/info:doi/   10.1371/journal.pone.0008767. Accessed 2013 Apr 3.

2. Hoorn C, Wesselingh FP, ter Steege H, Bermudez MA, Mora A, et al. (2010) Amazonia through time: Andean uplift, climate change, landscape evolution, and biodiversity. Science 330: 927–931. doi: 10.1126/science.1194585

3. Red Amazónica de Información Socioambiental Georreferenciada (2012) Mapa Amazonía 2012: Áreas Protegidas y Territorios Indígenas.

4. Killeen TJ (2007) A perfect storm in the Amazon wilderness: Development and conservation in the context of the Initiative for the Integration of the Regional Infrastructure of South America (IIRSA). Arlington, VA: Conservation International.

5. Finer M, Jenkins CN, Pimm SL, Keane B, Ross C (2008) Oil and Gas Projects in the Western Amazon: Threats to Wilderness, Biodiversity, and Indigenous Peoples. PLoS ONE 3(8): e2932. Available: http://www.plosone.org/article/info%3Ado i%2F10.1371%2Fjournal.pone.0002932. Accessed 2013 Apr 3.

6. Finer M, Moncel R, Jenkins CN (2010) Leaving the oil under the Amazon: Ecuador's Yasuní-ITT Initiative. Biotropica 42: 63–66. doi: 10.1111/j.1744-7429.2009.00587.x

7. Haselip J (2011) Transparency, consultation and conflict: Assessing the micro-level risks surrounding the drive to develop Peru's Amazonian oil and gas resources. Nat Resour Forum 35: 283–92. doi: 10.1111/j.1477-8947.2011.01404.x

8. Finer M, Orta-Martínez M (2010) Second hydrocarbon boom threatens the Peruvian Amazon: trends, projections, and policy implications. Environ. Res. Lett. 5: 014012. Available: http:// iopscience.iop.org/1748-9326/5/1/ 014012. Accessed 2013 Apr 3.

9. Benavides M (2009) Mapa Amazonia Peruana. Lima: Instituto del Bien Común. Available: http://www.ibcperu.org/mapas/mapa-

amazon ia-peruana.php. Accessed 2013 Apr 3.

10. Orta Martínez M, Napolitano DA, MacLennan GJ, O'Callaghan C, Ciborowski S, et al. (2007) Impacts of petroleum activities for the Achuar people of the Peruvian Amazon: summary of existing evidence and research gaps. Environ Res Lett 2: 1–10. doi: 10.1088/1748-9326/2/4/045006

11. Powers B (2012) Best Practices: Design of Oil and Gas Projects in Tropical Forests. E-Tech International.

12. República del Perú (2001) Ley del Sistema Nacional de Evaluación del Impacto Ambiental (Ley No. 27446), As modified by Decreto Legislativo No. 1078.

13. República del Perú; Reglamento de la Ley del SEIA (DS 019-2009-MINAM).

14. R. García Consultores SA, ARCAN Ingeniera y Construcciones SA, Centro de Conservación de Energía y del Ambiente (2012). Elaboración de la Nueva Matriz Energética Sostenible y Evaluación Ambiental Estratégica, como Instrumentos de Planificación.

15. Perupetro (2011) Marco general de las actividades de exploración y explotación de hidrocarburos.

16. Schlumberger (2011) Multi-property Earth model building through data integration for improved subsurface imaging. First Break 29.

17. National Petroleum Council (2011) Sustainable drilling of onshore oil and gas wells. Paper #2–23.

18. República del Perú (2006) Reglamento para la Protección Ambiental en las Actividades de Hidrocarburos. Decreto Supremo No. 015-2006-EM.

19. Tollefson J (2011) Fighting for the forest: The roadless warrior. Nature 480: 22–24. doi: 10.1038/480022a

20. Thurber M, Ayarza P (2005) Canopy bridges along a rainforest pipeline in Ecuador. Society of Petroleum Engineers paper SPE-96504-PP.

21. Ministerio de Energía y Minas del Perú (2007) Auditoría integral de los sistemas de transporte de gas natural y líquidos de gas natural del Proyecto Camisea. Reporte Final de Auditoria Integral.

22. The Energy & Biodiversity Initiative (2003) Integrating biodiversity conservation into oil and gas development. Washington DC: Conservation International.

23. Goodland R (2012) Responsible mining: the key to profitable resource development. Institute for Environmental Diplomacy and Security Research Series: A1-2012-4.

24. Fine PVA, Garcia-Villacorta R, Pitman NCA, Mesones I, Kembel SW (2010) A floristic study of the white-sand forests of Peru. Ann. Missouri Bot. Gard. 97: 283–305. doi: 10.3417/2008068

25. International Labour Organisation (1989) Convention No. 169 concerning Indigenous and Tribal Peoples in Independent Countries.

26. United Nations General Assembly (2007) United Nations Declaration on the Rights of Indigenous Peoples.

27. República del Perú (2011) Ley del Derecho a la Consulta Previa a los Pueblos Indígenas u Originarios, reconocido en el Convenio 169 de la Organización Internacional del Trabajo (OIT).

28. Zahniser A (2007) Characterization of greenhouse gas emissions involved in oil and gas exploration and production operations. Review for the California Air Resources Board.

29. Viale C (2011) Renta petrolera y uso del Canon en Loreto.

30. Viale C (2012) Generación, distribución y uso de la renta petrolera en Loreto. Lima: Programa de Vigilancia Ciudadana.

31. Ministerio de Energía y Minas del Perú (2011) Anuario estadístico de hidrocarburos 2011.

32. Petroperú (1981) Informe estadístico de petróleos del Perú 1970–1981.

33. Petroperú (1989) Informe estadístico annual 1989.

34. Ministerio de Energía y Minas del Perú (1999) Anuario estadístico de hidrocarburos 1999.

35. Ministerio de Energía y Minas del Perú (2012) Libro anual de reservas de hidrocarburos 2011.

36. Daimi Peru, Perenco (2010) EIA del proyecto de construcción del oleoducto y línea de diluyente CPF – Andoas.

37. Asamre Perenco (2011) Estudio de impacto ambiental para la fase de desarrollo del Lote 67A Y 67B.

38. Burlington Resources, ConocoPhillips (2012) Plan de abandono del proyecto de prospección sísmica 2D Lote123, Loreto.

39. Burlington Resources, ConocoPhillips (2012) Plan de abandono del proyecto de prospección sísmica 2D Lote129, Loreto.

40. Repsol GEMA (2011) EIA Proyecto de prospección sísmica 3D y perforación de 21 pozos exploratorios – Lote 39.

41. Ministerio de Energía y Minas del Perú (2012) Levantamiento de observaciones del EIA para el Proyecto Fase de Desarrollo de los Lotes 67A y 67B. Informe No. 001–2012-MEM-AAE/MMR.

42. Maple Energy (2012) Oil and gas exploration. Available: http://www.maple-energy.com/Oilmerger1.a spx. Accessed 2013 Apr 3.

43. Kerr RA (2010) Natural Gas From Shale Bursts Onto the Scene. Science 328: 1624–26. doi: 10.1126/science.328.5986.1624

44. Orta Martínez M, Finer M (2010) Oil frontiers and indigenous resistance in the Peruvian Amazon. Ecological Economics 70: 207–218. doi: 10.1016/j.ecolecon.2010.04.022

45. Pueblos Indígenas de Canaán de Cachiyacu y Nuevo Sucre (2010) Queja al Ombudsman de la CAO relativa a violaciones de derechos humanos y daños ambientales causados por Maple Energy.   Available:http://www.accountabilitycounsel.org/wp content/uploads/2011/12/Demanda-a-la-CAO Sobre-Maple.pdf. Accessed 2013 Apr 3.

46. Comisión de Pueblos Andinos, Amazónicos y Afroperuanos, Ambiente y Ecología (2012) Grupo de Trabajo sobre la Situación Indígena de las Cuencas de los Ríos Tigre, Pastaza, Corrientes y Marañón.

47. Thomsen JB, Mitchell C, Piland R, Donnaway JR (2001) Monitoring impact of hydrocarbon exploration in sensitive terrestrial ecosystems: perspectives from Block 78 in Peru. In: Bowles IA, Prickett GT, editors. Footprints in the jungle. New York: Oxford University Press.

48. Kolowski JM, Alonso A (2012) Primate abundance in an unhunted region of the northern Peruvian Amazon and the influence of seismic oil exploration. Int J Primatol 33: 958–971. doi: 10.1007/s10764-012-9627-y

49. Kolowski JM, Alonso A (2010) Density and activity patterns of ocelots (*Leopardus pardalis*) in northern Peru and the impact of oil exploration activities. Biol Conserv 143: 917–25. doi: 10.1016/j.biocon.2009.12.039

50. Orpio Corpi-Sl, Fediquep Acodecospat, Feconat, et al. (2012) Nuestras condiciones previas para poder iniciar el proceso de consulta previo.

51. Perupetro (2011) Plays and new hydrocarbon potential in the Marañon Basin.

52. Perupetro (2012) Next bidding round 2012: 36 blocks for exploration activities.

53. Lehner B, Verdin K, Jarvis A (2008) New global hydrography derived from spaceborne elevation data. Eos Trans AGU 89: 93–94. doi: 10.1029/2008eo100001

54. SERNANP (Servico Nacional de Áreas Naturales Protegidas por el Estado) (2009) Plan Director de las Áreas Naturales Protegidas (Estrategia Nacional). Ministerio del Ambiente.

55. Instituto del Bien Común (2011) Sistema de Información sobre Comunidades Nativas de la Amazonía Peruana (SICNA), versión 4. Lima: IBC.

56. Vriesendorp C, Pitman N, Rojas Moscoso JI, Pawlak BA, Chávez LR, et al.. (2006) Perú: Matsés. Rapid Biological Inventories Report 16. Chicago: The Field Museum.

57. Josse C, Navarro G, Encarnación F, Tovar A, Comer P, et al.. (2007) Digital ecological systems map of the Amazon Basin of Peru and Bolivia. Arlington: NatureServe.

58. Mims M (2002) Drilling design and implementation for extended reach and complex wells. KM Technology Group.

59. Lee J (2009) Introduction to offshore pipelines and risers.

60. Amores G (2010) Comparaciones de calidad y costo entre un gasoducto verde y una construcción tradicional – el control de la erosión como medida de protección ambiental. INMAC Perú.

61. Olson DM, Dinerstein E, Wikramanayake ED, Burgess ND, Powell GVN, et al. (2001) Terrestrial ecoregions of the world: a new map of life on Earth. BioScience 51: 933–938. doi: 10.1641/0006-3568(2001)051[0933:teotwa]2.0.co;2

Perceptual Characterization and Analysis of Aroma Mixtures Using Gas Chromatography Recomposition-Olfactometry

Arielle J. Johnson, Gregory D. Hirson, and Susan E. Ebeler

Department of Viticulture and Enology, Agricultural and Environmental Chemistry Graduate Group, University of California Davis, Davis, California, United States of America

ABSTRACT

This paper describes the design of a new instrumental technique, Gas Chromatography Recomposition-Olfactometry (GC-R), that adapts the reconstitution technique used in flavor chemistry studies by extracting volatiles from a sample by headspace solid-phase microextraction (SPME), separating the extract on a capillary GC column, and recombining individual compounds selectively as they elute off of the column into a mixture for sensory analysis (Figure 1). Using the chromatogram of a mixture as a map, the GC-R instrument allows the operator to "cut apart" and recombine the components of the mixture at will, selecting compounds, peaks, or sections based on retention time to include or exclude in a reconstitution for sensory analysis. Selective recombination is accomplished with the installation of a Deans Switch directly in-line with the column, which directs compounds either to waste or to a cryotrap at the operator's discretion. This enables the creation of, for example, aroma reconstitutions incorporating all of the volatiles in a sample, including instrumentally undetectable compounds as well those present at concentrations below sensory thresholds, thus correcting for the "reconstitution discrepancy" sometimes noted in flavor chemistry studies. Using only flowering lavender (*Lavandula angustifola* 'Hidcote Blue') as a source for volatiles, we used the instrument to build mixtures of subsets of lavender volatiles in-instrument and characterized their aroma qualities with a sensory panel. We showed evidence of additive, masking, and synergistic effects in these mixtures and of "lavender› aroma character as an emergent property of specific mixtures. This was accomplished without the need for chemical standards, reductive aroma models, or calculation of Odor Activity Values, and is broadly applicable to any aroma or flavor.

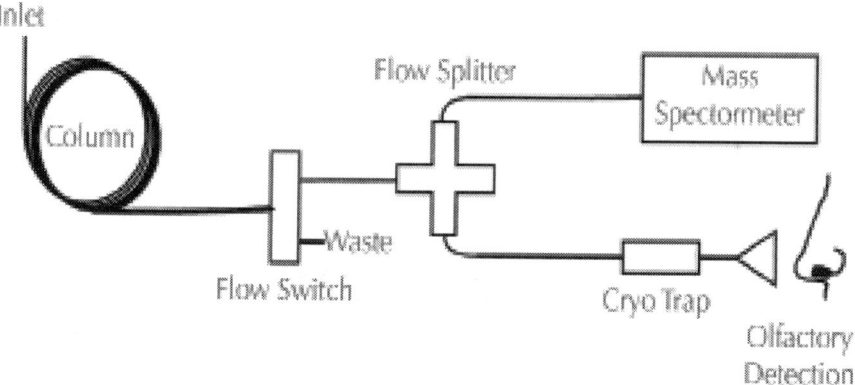

Figure 1: Conceptual schematic of the Gas Chromatograph Recomposition-Olfactometer (GC-R) instrument. Volatiles are extracted onto a solid phase (via solid-phase microextraction or SPME) from the headspace of a food, bev-

erage, or other sample, in this case, lavender flowers, and initially they are separated conventionally on an analytical capillary GC column. In-line with the GC column, a pneumatic Deans Switch followed by a cold trap allows the experimenter to build a mixture of these separated volatiles that is held until the cryotrap is rapidly heated, releasing the mixture for a subject to smell at the olfactory port and evaluate.

INTRODUCTION

Aroma plays a dominant role in the multisensory perception of flavor. It is itself a construct perceived in response to stimulation of the olfactory system by volatile chemicals and mixtures thereof, with mixtures being commonly encountered in everyday life in the form of food, wine, plants, perfume, etc. While our understanding of the neurobiological and psychological mechanisms that translate volatiles into aroma perceptions has advanced significantly in recent years [1], [2], analytical approaches for characterizing the perception of these aroma mixtures are still limited. The relationship between chemical composition of a mixture of volatiles and its perceived aroma or flavor is complex and difficult to predict on the basis of chemical data or simple sensory data alone.

Analytical chemistry approaches for characterizing aromas or flavors typically rely on separation-based chromatographic methods that quantify the aroma strength of individual compounds in a mixture, reflected as either the concentration present in the mixture divided by a measured sensory threshold concentration (Odor Activity Value, OAV) [3], [4] or the number of N-fold dilutions required to suppress detectability of a compound when analyzed by gas chromatography with a human subject acting as an olfactory detector (GC-Olfactometry or GC-O; CHARM; or Aroma Extract Dilution Analysis) [5]–[7]. Reconstitution and omission experiments evaluate the role of specific compounds in the perceived aroma of a mixture, whereby a blend of compounds hypothesized to be detectable in a food, beverage, or other sample by OAV is mixed from chemical standards, and compared to similar mixtures prepared by omitting one of these compounds at a time [7]. If a difference is detectable in the "whole" mix versus a "whole-minus-one-compound" mix, that particular compound is considered important to the aroma of the sample.

Knowledge from other disciplines studying aroma, such as sensory psychophysics, cognitive psychology, and molecular neurobiology, suggests limitations of these methodologies. Chromatographic techniques only assess the aroma quality of individual compounds, rather than mixtures of compounds. However, the aroma of a mixture is frequently perceptually distinct from that of its individual components [8], [9] and may have qualities not found in any of these components [10]. The mixing-dependent nature of aroma quality is evidenced by the relative lack of aroma impact compounds, or those compounds that are singularly responsible for the overall aroma impression of a food or beverage. On the other hand, omission experiments rely on an assumption that all sensorially important compounds have been correctly identified and quantified and that any compound occurring at a concentration below its putative sensory threshold is not important to the overall aroma. Recently published results suggest that this is not the case [11]. Despite having identical concentration profiles of supra-threshold odorants, the aroma of a reconstitution sometimes still smells different from the original mixture [12], a phenomenon referred to as "reconstitution discrepancy" [13]. Some recent omission experiments have included sub-threshold components in the reconstitution[13], but this is not a universal practice, and can greatly complicate and enlarge the experimental design.

We propose here a novel platform for the analytical characterization of aroma and flavor perception that incorporates and merges aspects of the previously described techniques and knowledge from other related disciplines. We describe a series of non-reductive, in-instrument recombination and omission experiments using a Gas Chromatograph modified with a switch and then a cold trap in-line between the capillary column and the chemical and olfactory detectors to characterize the aroma of lavender (Lavandula angustifola 'Hidcote Blue'). The volatile chemical composition of lavender, a potently aromatic herb with numerous culinary, cosmetic, and fragrance uses, has previously been characterized [14], but there are no lavender impact compounds currently identified. This suggests that "lavender" aroma character arises from the perception of a mixture of volatiles rather than a single molecule, making this an ideal mixture for evaluation of perceptual interactions using our gas chromatography recomposition-olfactometry GC-R) approach.

MATERIALS AND METHODS

Instrument

An Agilent model 6890 gas chromatograph/5972 mass spectral detector (GC-MSD) was modified with the addition of a Deans switch apparatus (Agilent Technologies, Santa Clara, CA), an auxiliary pressure controller (EPC, Agilent) to control flow through the Deans switch, a splitter (Gerstel), a cryotrap (Micro Cryo-trap and model 971 controller, Scientific Instrument Services, Ringoes NJ) and an olfactometry port (ODP-2, Gerstel, Linthicum, MD). A schematic showing modifications from a standard GC-MS (Figure 2a), to a GC-O instrument (Figure 2c), to the GC-R Gas Chromatograph is shown (Figure 2c). Deactivated fused silica was used for all transfer lines. The transfer line from the Deans switch to the splitter was 4 m. The dimensions of the transfer line from the splitter to the MSD was 1 m × 0.15 mm; the dimensions of the transfer line from the splitter to the olfactory port was 1 m × 0.25 mm resulting in a 1.86:1 split ratio between the olfactory port and MSD.

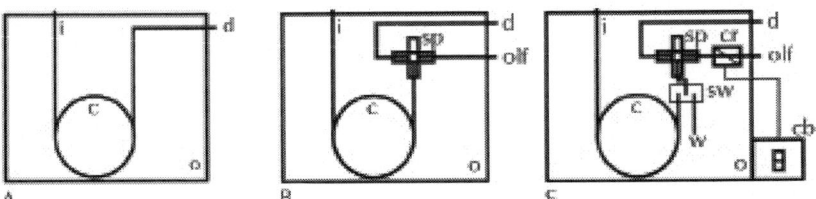

Figure 2: Schematic of (a) standard GC-MS; (b) GC-MS with splitter at end of column for olfactometry; and (c) Gas Chromatograph-Recomposition-Olfactometer or GC-R with Deans switch, splitter, cryogenic trap and olfactory port. Abbreviations: i-inlet; c-column; d-detector; o-oven; olf-olfactometry port; sp-splitter; sw-Deans switch 1; w-waste; cr-cryogenic trap; and cb-switch 2 on control box.

Sampling and Chromatographic Conditions

Lavender (*Lavandula angustifola* 'Hidcote Blue') flowers (0.50 g) were weighed and placed in a 20 mL amber glass headspace vial and sealed with a crimp cap with a PTFE-faced silicone septum (Supelco, St. Louis, MO). A Solid Phase Microextraction fiber (2 cm length, 50/30 um divinylbenzene/carboxen/polydimethylsiloxanecoating, Supelco) was used for extraction. The fiber was exposed to the headspace of the vial for 30 minutes at room temperature, then withdrawn and immediately desorbed in the GC inlet. Chromatographic conditions were adapted from [14]. Separation was performed using a 30 m×25 mm i.d. ×0.25 um film thickness DB-5MS column (J&W, Folsom, CA). The inlet was maintained at 240°C in splitless mode. Helium was used as the carrier gas and was held at constant pressure at 15.5 psi. The auxiliary pressure controller was maintained at 3.4 psi. The SPME assembly was introduced manually into the inlet and allowed to desorb for a total of 10 minutes. The oven was held at 60°C for 3 minutes, then ramped to 150°C at a rate of 3°C/min, then ramped to 325 at a rate of 30°C/min and held for 1 min for a total runtime of 40 minutes. The olfactory port transfer line was maintained at 100°C and the MSD transfer line was maintained at 260°C. After a 0.5 min solvent delay, the mass spectrometer scanned from m/z 50–230. With the Deans switch set in the "off" position, the flow is directed to the splitter, MSD, cold trap, and ODP. When set to the "on" position, the flow is directed to waste. The switch is programmed in the "runtime" tab of the Enhanced Chemstation Software (Hewlett Packard, version B.01.00) to direct the flow over the course of the runtime as desired by the operator.

Sensory Conditions

Based on retention time, the Deans Switch sends specific packets of volatiles to the cryotrap. Here we used one of ten programs (W, O1–O3, P1–P6; see Figure 3, Table 1) where at the conclusion of the separation run, the cold trap was heated and the mixture was sniffed and described by a sensory panelist. The W condition, analogous to a full aroma reconstitute, contains all the volatiles of lavender, with conditions O1–O3 and P1–P6 omitting groups of these volatiles for descriptive comparison to the aroma of the W sample and to lavender flowers.

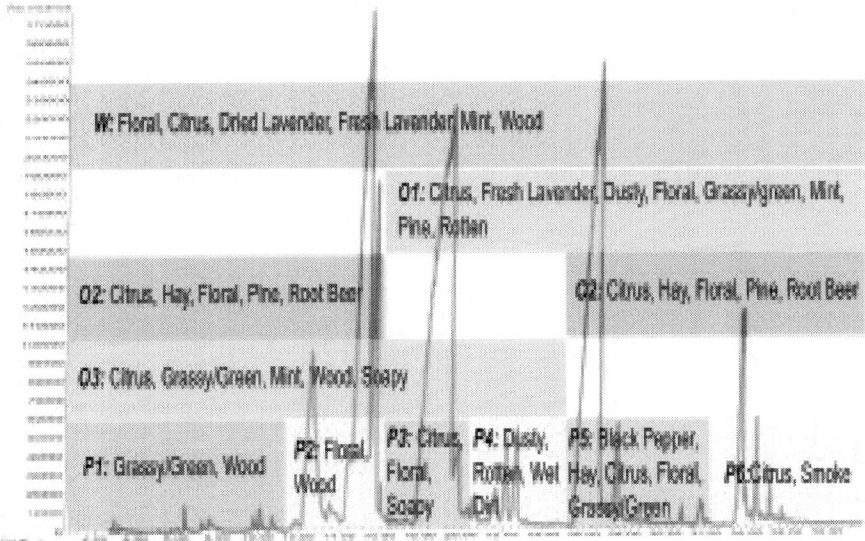

Within the figure:

W: Floral, Citrus, Dried Lavender, Fresh Lavender, Mint, Wood

O1: Citrus, Fresh Lavender, Dusty, Floral, Grassy/green, Mint, Pine, Rotten

O2: Citrus, Hay, Floral, Pine, Root Beer

O2: Citrus, Hay, Floral, Pine, Root Beer

O3: Citrus, Grassy/Green, Mint, Wood, Soapy

P1: Grassy/Green, Wood   P2: Floral, Wood   P3: Citrus, Floral, Soapy   P4: Dusty, Rotten, Dirt   P5: Black Pepper, Wet Hay, Citrus, Floral, Grassy/Green   P6: Citrus, Smoke

Figure 3: Top aroma descriptors for mixtures of sections of the lavender chromatogram by cut time and chromatogram composition. Abbreviations correspond to Experimental Conditions described in Table 1. As chemical complexity and number of components per mixture approaches the makeup of the whole chromatogram (W) mixture, there is evidence of perceptual additivity as increasing cross-utilization of terms from simpler mixtures, masking as reduced use of dominant terms for simpler (P1–P6) mixtures, and synergistic effects as new complex or composite terms like "fresh lavender" become important.

Table 1: Experimental GC-O conditions and aroma descriptors for mixtures of volatiles from the lavender chromatograms

Experimental Condition	Abbreviation	Chromatogram Sections Included in Mixture	Top Descriptors
Whole Chromatogram	W	0-40 minutes	Floral, citrus, dried lavender, fresh lavender, mint, wood

'Omission 1	01	16-40 minutes	Citrus, fresh lavender, dusty, floral, grassy/green, mint, pine, rotten
'Omission 2	02	0-16+25-40 minutes	Citrus, haylike, floral, pine, root beer
'Omission 3	03	0-25 minutes	Citrus, grassy/green, mint, wood, soapy
2Perceptual Interaction 1	P1	0-11 minutes	Grassy/green, wood
2Perceptual Interaction 2	P2	11-16 minutes	Floral, wood
2Perceptual Interaction 3	P3	16-20.5 minutes	Citrus, floral, soapy
2Perceptual Interaction 4	P4	20.5-25 minutes	Dusty, rotten, wet dirt
2Perceptual Interaction 5	P5	25-32 minutes	Black pepper, haylike, citrus, floral, grassy/green
2Perceptual Interaction 6	P6	32-40 minutes	Citrus, smoke
Lavender Flowers Reference	Reference	Not separated; whole lavender flowers	Citrus, floral, fresh lavender, mint, wood, hay, dried lavender, grassy/green

Three panelists (Females, ages 28–45 with previous sensory experience) smelled each of the ten mixtures in triplicate and generated terms to describe the perceived odor. Before smelling each mixture, each panelist first smelled and described a standard of lavender flowers, picked at the same time as the flowers used for SPME sampling, and also rated how well the sample mixture represented the aroma of the standard on a scale of 0–10.

Ethics Statement

Use of human subjects for this study was reviewed by the University of California, Davis Institutional Review Board and was granted exempt status (Category 6).

Data Analysis

The terms used to describe the ten mixtures were tabulated by frequency of use. The descriptors used most often for each mixture, in a mixture-by-descriptor data matrix, were analyzed with a correspondence analysis to identify latent trends in similarity and difference in the multidimensional set. A three-way Analysis of Variance (ANOVA) with all two-way interactions was performed; rated representativeness of each mixture was compared to a fresh lavender standard as the response factor and panelist, mixture, and replicate were the main effects. A Tukey's Honest Significant Difference multiple comparisons test (HSD) was performed on the representativeness ratings. The R statistical computing package was used for all statistical analyses (http://www.r-project.org/).

RESULTS AND DISCUSSION

We modified a GC-MS to allow for the in-instrument preparation of volatile mixtures containing precise sections from a chromatogram, up to and including the entire volatile fraction and allowing for aroma characterization of the aroma of one or a few of the volatiles in a complex mixture (Figure 1). Compounds were introduced into the inlet of the modified GC-MS and separated on the analytical column. At the end of the column, the flow of carrier gas and analytes encountered a first switch, a commercially available Deans switch, that was set to direct the flow either towards the splitter or towards waste (here waste was vented to the oven). The splitter subsequently split the flow to both a mass spectrometer (MS) detector and to an olfactory port. Along the transfer line to the olfactory port was a trap controlled by a second switch at the control box; the switch allowed the trap to be cooled with liquid carbon dioxide or heated so that the eluant was either held within the trap (*i.e.*, cryotrapped) or released to the olfactory port. By programming the switches to cryotrap or exclude selected peaks or peak regions (Table 1) two types of experiments were performed. In perceptual interaction experiments, all of the chromatogram except for a small section of peaks was cut away, and the section of interest was assessed at the olfactory port as a mixture. In omission experiments, small groups of peaks (or individual peaks) were cut away and the rest of the compounds in the chromatogram were smelled as a mixture. See Figures S1 and S2 for examples of these chromatograms.

Using our new approach, ten aroma mixtures (Table 1, Figure 3) were created in-instrument directly from the headspace-extracted volatiles of flowering lavender. "Fresh Lavender" and "Dried Lavender" were both predominant descriptors for the "Whole Volatile" recombination mixture W. Of the more chemically complex omission mixtures O1–O3, only O1, which incorporated the section of volatiles eluting from 16–40 min of the lavender chromatogram and omitted volatiles eluting between 0–16 min, was described as having "fresh lavender" properties. O1 overlapped with O2 from 25–40 min and with O3 from 16–25 minutes and incorporated the perceptual mixtures P3–P6, however, none of these other omission or perceptual mixtures had fresh or dried lavender among their commonly used descriptors. This suggests that there are two subsets of compounds, the first eluting between 16–25 min and the other eluting between 25–40 min, that are each necessary for the perception of "lavender character" but are not alone sufficient for inducing this perception without some mixing with compounds in the other elution group. These results also suggest that "lavender character" is an emergent perceptual property arising from the mixing of these volatiles or some subset thereof.

We performed a Correspondence Analysis on the descriptors-by-mixtures data matrix to compare dimensionally-reduced latent trends in the sensory profiles of the mixtures to the differences evident in top descriptors for each mixture (Figure 4). Correspondence Analysis separates dissimilar categories in space; mixtures and sensory descriptors spaced closely together share more similarities than those spaced further apart. This plot shows that, generally, removing more volatiles results in greater dissimilarity between a given mixture and the all-volatiles-included mixture W. The relatively tight clustering of W and omission mixtures O1–O3 in the Correspondence Analysis reflects the sensory similarity of these mixtures; perceptual mixtures P2 and P3 also cluster nearby, reflecting some of the overlapping characteristics of these mixtures (Figure 4).

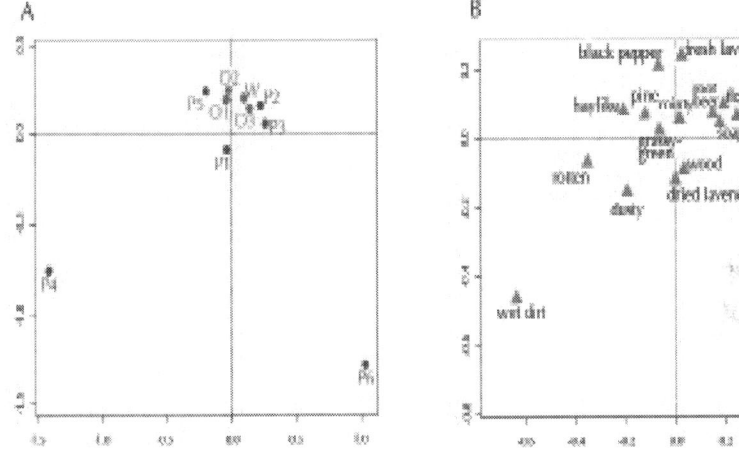

Figure 4: Correspondence Analysis of (A) lavender volatile mixtures; and (B) lavender volatile mixture descriptors. Abbreviations for mixtures correspond to those in Table 1. Terms generated by the panelists to describe the perceived odor of from each Experimental Condition described in Table 1 were tabulated by frequency of use and used for the Correspondence Analysis. 30.57% of variance explained by dimension 1 (x), 22.84% of variance explained by dimension 2 (y).

The location of mixture W in the center of the main cluster in the Correspondence Analysis, suggests its aroma was perceived, in part, as a sensory average of some of the less-complex mixtures. However, a truly averaged perceptual character would be in the center of the plot; the fact that mixture W is offset from the geometric center implies that the mixing-dependent interactive effects of the lavender volatiles perceived in mixture W play a noticeable role in affecting its overall aroma character. Mixture W shares many similar descriptors (Table 1) with O1–O3 and P2 and P3, but all of these except O1 lack a dominant lavender character.

Mixtures P1 and P5 are close to the central cluster but are approximately equi-distant in space from mixture W. This reflects some of the similarities in the descriptors that P1 and P5 share with mixture W, but also reflects the domination of the aromas of these mixtures by either a unique character ("black pepper") in the case of P5, or the relative simplicity of the aroma in the case of P1 (Figure 4a). The comparative distancing of mixtures P4 and P6 from the other mixtures reflects the relative uniqueness of their aroma descriptors.

Locations of descriptors suggest that along the first (x) dimension of Figure 4b, there is a distinction between fresher, more "sweet" and flower-associated terms on the right side and earthier, heavier aroma terms on the left. Borrowing more qualitative terms from the tradition of perfumery (which at its essence is the craft of observing and optimizing the perceptual effects of mixing volatiles), we observe a rough progression, from left to right along the x-axis, of base, middle, and top-note [15] related terms. Along the second (y) dimension the separation is dominated by the marked difference of P4 and P6 from each other and from the rest of the mixtures, and correspondingly by their unique descriptors "wet dirt" and "smoky" in Figure 4b. Generally, the terms on the other arm of the y-dimension tend to be shared by multiple mixtures, or reflect more composite aroma characteristics.

While sample P1 appears to be the closest to the central or average sample in this set, it is clearly separated from the cluster centered around mixture W along the third (z) dimension. The third dimension also further separates mixture P5 from the central W-associated cluster and increases the distinction between "grassy/green"-"woody" descriptors on one side and "dried lavender"-"black pepper" descriptors on the other. Importantly, the Correspondence Analysis, while unable to describe absolute differences, provides valuable information not only on the sources of variation in the complex sensory data but also on the interrelationships of the mixtures and their sensory properties.

The method used to create an extract of volatile compounds can alter the perceived aroma of that extract and failure to obtain a representative sample can lead to unreliable conclusions about the composition of the aroma active components [16]–[20]. While many extraction methods have been employed in order to produce an aroma extract [19]–[21], the creation of a representative aroma can be very difficult for complex matrices [20], [22], and the sensory representativeness of this extract is not always evaluated. Here, the aroma of the SPME extracts of lavender corresponded closely to the original product (Table 1). Similar representative aroma samples have been obtained using SPME to sample "baked potato" aroma [23]. Importantly, the GC-R approach provides a rapid, easy, and effective tool to assess the representativeness of an extract regardless of the extraction method employed, such as in cases where SPME coatings may not be able to produce an appropriate extract [24].

Since the SPME extraction produced an aroma mixture representative of lavender, it was possible to perform omission and interaction experiments based on a starting point nearly identical to the intact lavender sample, eliminating "reconstitution discrepancy" [13]. Comparing the aroma of the GC-R mixtures in this study to the aroma of whole lavender flowers, panelists found that mixtures P1, P5, and P6 were significantly less representative (Figure 5) of the aroma of the whole flowers than mixtures W, O1–O3 and P2–P4. These samples also tended to have either fewer commonly used descriptors or descriptors not found for other mixtures (such as "black pepper" for P5 and "smoke" for P6; Table 1).

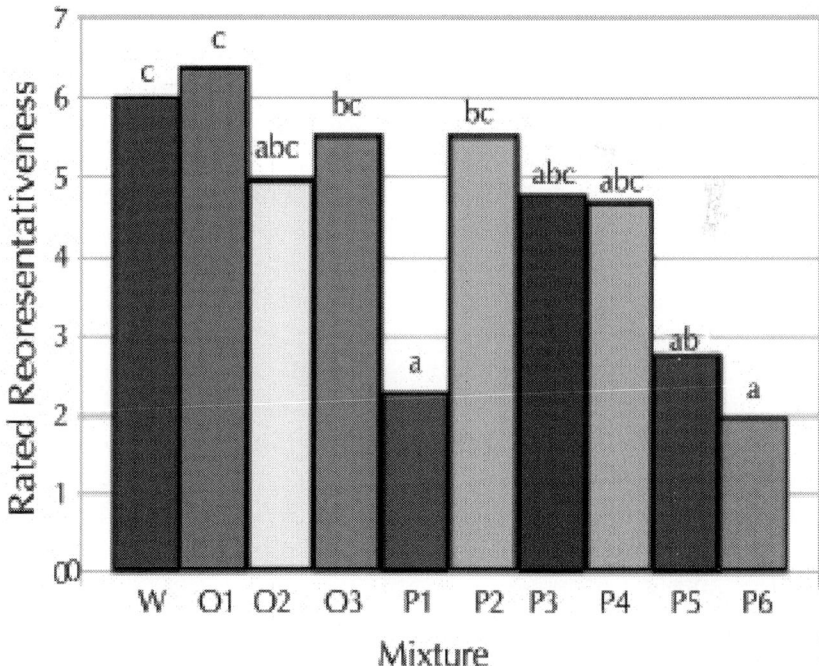

Figure 5: The rated representativeness of the aroma of samples W, O1–O3, and P1–P6 as compared by panelists to the aroma of whole flowering lavender. Letters a, b, c refer to the mixture's Significant Difference from each other- if two samples do not share a letter, they are significantly different. Samples P1, P5, and P6 are significantly less representative of the aroma of flowering lavender than sample W, which incorporates all the volatiles in flowering lavender.

In this experimental design, mixtures of compounds were omitted to assess the resulting aroma. Cut times were chosen to include chemically similar compounds in the same mixture, for example, monoterpene acetate esters in mixture P5 and sesquiterpenes in mixture P6. However, the omitted compounds/fractions in a theoretical GC-R experiment need not be contiguous. It is possible, for example, to remove every other chromatographic peak, to remove only the 3rd and 17th peak, etc. while trapping and evaluating the remaining components. The apparatus could additionally be used to perform single omission experiments, where compounds are omitted one at a time to screen for potential impact odorants, or perceptual interaction experiments where only 2 or 3 peaks are included in the mixture. The flexibility in the compounds that can be removed and assessed is only limited by the rapid switching time of the Deans switch. By using a Mass Spectrometric detector, compounds in the sample can be identified however, an obvious advantage of performing an omission experiment in this manner is that the compounds need not be identifiable or available to perform the experiment. Reconstitution experiments often require the experimenter to perform lengthy and labor-intensive syntheses to prepare a component for the reconstitution model [12] only to find that the component can be omitted with no change in the overall aroma of the solution. Furthermore, there is always some fraction of the total compounds identified that are not included in the reconstitution because they are deemed to have a concentration too low to have an effect on the overall aroma. However, compounds with low odor activity values often still have a considerable effect on the overall aroma of the mixture [11], [25], [26]. With this instrument there is no simplified reconstitute - the omission experiment is performed on the entire sample.

While compounds with low OAVs may be important to the aroma of the mixtures, the opposite case can also occur, and the sensitivity of the human nose is frequently orders of magnitude greater than an instrumental detector. As a result, the nose may detect an aroma where there is no peak on a chromatogram [17]. Particularly as compared to reconstitution studies, this is another distinct advantage of the GC-R approach since even compounds not detected by the detector (MS, FID) will be included in the aroma sample as it is assessed by a subject at the olfactometry port. Traditionally, full separation of volatile compounds on the chromatographic column is necessary in order to meaningfully describe the aroma character of the eluant by GC-O since it simplifies the recognition task for the assessor [21]. However, it is more often the case that a complex mixture of aroma compounds is responsible for the overall aroma of a food or beverage. In addition, a mixture of two or more odorants can frequently lead to an aroma that is not similar to any of its individual components [10], [27]. Using a GC-R technique, any of these interactions can readily be investigated; and all that is necessary to characterize any type of aroma interaction is a sample of the food, beverage, flower, etc. of interest. Compounds detectable by GC-O but not GC-MS, compounds below putative aroma thresholds, compounds at levels that cannot be quantified, and compounds not commercially available or easily synthesized can all be perceptually analyzed if they are found in one or more aromatic samples available to the researcher.

CONCLUSIONS

The perception of aroma and flavor has often been approached as a problem of many individual parts, with chemistry, neurobiology, sensory science, psychology, and other disciplines focused on answering questions about some aspect of the relationship between stimulus (a flower, a glass of wine, a plate of food), response (perceived flavor, liking or disliking, intake and satiety), or the pathway between the two (genetics, receptor binding, transduction, translation to cortical neurons). This has yielded a great deal of information about those individual parts, but not a well-developed understanding of how they work together for complex, everyday stimuli and activities like eating and drinking. The need for a holistic approach to address this has been identified previously [28], i.e., a praxis which would bring together knowledge and research techniques from these diverse, often isolated, but orthogonally-related scientific fields, and would include expertise or information from applied, non-analytical fields with a well-developed shared intuition about the nature of aroma and flavor in practice, such as cuisine and perfumery. While the described approach of in-instrument gas chromatography recombination-olfactometry has its roots in a traditional coupling of analytical chemistry and sensory science, it is highly informed by this multidisciplinary understanding of aroma and flavor and allows for the analysis of previously uncharacterized emergent perceptual properties of complex mixture interaction effects in everyday smell and flavor situations.

ACKNOWLEDGMENTS

We thank Jim McCurry and Agilent Technologies for the auxiliary pressure controller and John Thorngate for helpful discussions in early stages of this project. We thank the sensory judges who participated in this study.

AUTHOR CONTRIBUTIONS

Conceived and designed the experiments: AJJ GDH SEE. Performed the experiments: AJJ GDH. Analyzed the data: AJJ. Contributed reagents/materials/analysis tools: AJJ GDH. Wrote the paper: AJJ GDH SEE.

REFERENCES

1.   Buck LB (2004) Unraveling the sense of smell. Les Prix Nobel. The Nobel Prizes 2004. pp. 267–283.

2.   Axel R (2004) Scents and sensibility: A molecular logic. Les Prix Nobel. The Nobel Prizes 2004. pp. 234–256.

3.   Patton S, Josephson D (1957) A method for determining significance of volatile flavor compounds in foods. J Food Sci 22: 316–318. doi: 10.1111/j.1365-2621.1957.tb17017.x

4.   Guadagni DG, Buttery RG, Harris J (1966) Odour intensities of hop oil components. J Sci Food Agric 17: 142–144. doi: 10.1002/jsfa.2740170311

5.   Acree TE, Barnard J (1984) A procedure for the sensory analysis of gas chromatographic effluents. Food Chem 14: 273–286. doi: 10.1016/0308-8146(84)90082-7

6.   Grosch W (1993) Detection of potent odorants in foods by aroma extract dilution analysis. Trends Food Sci Techn 4: 68–73. doi: 10.1016/0924-2244(93)90187-f

7. Grosch W (2001) Evaluation of the key odorants of foods by dilution experiments, aroma models and omission. Chem Sens 26: 533–545. doi: 10.1093/chemse/26.5.533

8. Wilson DA, Stevenson RJ (2006) Learning to smell: olfactory perception from neurobiology to behavior. Baltimore: JHU Press. 297 p.

9. Laing DG, Francis GW (1989) The capacity of humans to identify odors in mixtures. Physiol Behav 46: 809–814. doi: 10.1016/0031-9384(89)90041-3

10. Le Berre E, Béno N, Ishii A, Chabanet C, Etiévant P, et al. (2008) Just noticeable differences in component concentrations modify the odor quality of a blending mixture. Chem Senses 33: 389–395 doi:10.1093/chemse/bjn006. doi: 10.1093/chemse/bjn006

11. Pineau B, Barbe J-C, Van Leeuwen C, Dubourdieu D (2007) Which impact for beta-damascenone on red wines aroma? J Agric Food Chem 55: 4103–4108 doi:10.1021/jf070120r. doi: 10.1021/jf070120r

12. Steinhaus M, Sinuco D, Polster J, Osorio C, Schieberle P (2009) Characterization of the key aroma compounds in pink guava (*Psidium guajava* L.) by means of aroma re-engineering experiments and omission tests. J Agric Food Chem 57: 2882–2888 doi:10.1021/jf803728n. doi: 10.1021/jf803728n

13. Bult JH, Schifferstein HN, Roozen JP, Voragen AG, Kroeze JH (2001) The influence of olfactory concept on the probability of detecting sub- and peri-threshold components in a mixture of odorants. Chem Senses 26: 459–469. doi: 10.1093/chemse/26.5.459

14. An M, Hai T, Hatfield P (2001) On-site field sampling and analysis of fragrance from living lavender (*Lavandula angustifolia* L.) flowers by solid-phase microextraction coupled to gas chromatography and ion-trap mass spectrometry. J Chromatogr A 917: 245–250. doi: 10.1016/s0021-9673(01)00657-4

15. Aftel M (2004) Essence and alchemy: A natural history of perfume. Layton, UT: Gibbs Smith. 244 p.

16. Abbott N, Etievant P, Langlois D, Lesschaeve I, Issanchou S, et al. (1993) Evaluation of the representativeness of the odor of beer extracts prior to analysis by GC eluate sniffing. J Agric Food Chem 41: 777–780. doi: 10.1021/jf00029a019

17. Etiévant PX, Moio L, Guichard E, Langlois D, Leschaeve I, et al.. (1993) Aroma extract dilution analysis (AEDA) and the representativeness of the odour of food extracts. In: Maarse H, Van Der Heij DG, editors. Trends in flavour research, volume 35 of Developments in food science. New York: Elsevier. pp. 179–190.

18. van Ruth SM, Geary MD, Buhr K, Delahunty CM (2004) Representative sampling of volatile flavor compounds: The model mouth combines with gas chromatography and direct mass spectrometry. In: Deibler KD, Delwiche J, editors. Handbook of flavor characterization: Sensory analysis, chemistry, and physiology. New York: Marcel Dekker, Inc. pp. 303–311.

19. Plutowska B, Wardencki W (2008) Application of gas chromatography–olfactometry (GC–O) in analysis and quality assessment of alcoholic beverages – A review. Food Chem 107: 449–463 doi:10.1016/j.foodchem.2007.08.058. doi: 10.1016/j.foodchem.2007.08.058

20. Aceña L, Vera L, Guasch J, Busto O, Mestres M (2010) Comparative study of two extraction techniques to obtain representative aroma extracts for being analysed by gas chromatography-olfactometry: application to roasted pistachio aroma. J Chromatogr A 1217: 7781–7787 doi:10.1016/j.chroma.2010.10.030. doi: 10.1016/j.chroma.2010.10.030

21. San-Juan F, Pet'ka J, Cacho J, Ferreira V, Escudero A (2010) Producing headspace extracts for the gas chromatography–olfactometric evaluation of wine aroma. Food Chem 123: 188–195 doi:10.1016/j.foodchem.2010.03.129. doi: 10.1016/j.foodchem.2010.03.129

22. Pérez-Silva A, Odoux E, Brat P, Ribeyre F, Rodriguez-Jimenes G, et al. (2006) GC–MS and GC–olfactometry analysis of aroma compounds in a representative organic aroma extract from cured vanilla (Vanilla planifolia G. Jackson) beans. Food Chem 99: 728–735 doi:10.1016/j.foodchem.2005.08.050. doi: 10.1016/j.foodchem.2005.08.050

23. Poinot P, Grua-Priol J, Arvisenet G, Rannou C, Semenou M, et al. (2007) Optimisation of HS-SPME to study representativeness of partially baked bread odorant extracts. Food Res Internat 40: 1170–1184 doi:10.1016/j.foodres.2007.06.011. doi: 10.1016/j.foodres.2007.06.011

24. Ferreira V, Lopez R, Aznar M (2002) Olfactometry and aroma extract dilution analysis of wines. In: Jackson JF, Linskens HF, editors. Analysis of taste and aroma. Volume 21. Heidelberg: Springer-Verlag. pp. 89–122.

25. Escudero A, Gogorza B, Melus MA, Ortin N, Cacho J, et al. (2004) Characterization of the aroma of a wine from maccabeo. Key role played by compounds with low odor activity values. J Agric Food Chem 52: 3516–3524 doi:10.1021/jf035341l. doi: 10.1021/jf035341l

26. Ryan D, Prenzler PD, Saliba AJ, Scollary GR (2008) The significance of low impact odorants in global odour perception. Trends Food Sci Techn 19: 383–389 doi:10.1016/j.tifs.2008.01.007. doi: 10.1016/j.tifs.2008.01.007

27. Le Berre E, Thomas-Danguin T, Béno N, Coureaud G, Etiévant P, et al. (2008) Perceptual processing strategy and exposure influence the perception of odor mixtures. Chem Senses 33: 193–199 doi:10.1093/chemse/bjm080. doi: 10.1093/chemse/bjm080

28. Shepherd GM (2006) Smell images and the flavour system in the human brain. Nature 444: 316–321 doi:10.1038/nature05405. doi: 10.1038/nature05405

Preparation and Layer-by-Layer Solution Deposition of Cu(In,Ga)O$_2$ Nanoparticles with Conversion to Cu(In,Ga)S$_2$ Films

Walter J. Dressick[1], Carissa M. Soto[1], Jake Fontana[1], Colin C. Baker[2], Jason D. Myers[2], Jesse A. Frantz[2], and Woohong Kim[2]

[1]Center for Bio/Molecular Science & Engineering, United States Naval Research Laboratory, Washington, District of Columbia, United States of America,

[2]Optical Sciences Division, United States Naval Research Laboratory, Washington, District of Columbia, United States of America

ABSTRACT

We present a method of Cu (In,Ga)S$_2$ (CIGS) thin film formation via conversion of layer-by-layer (LbL) assembled Cu-In-Ga oxide (CIGO)

nanoparticles and polyelectrolytes. CIGO nanoparticles were created via a novel flame-spray pyrolysis method using metal nitrate precursors, subsequently coated with polyallylamine (PAH), and dispersed in aqueous solution. Multilayer films were assembled by alternately dipping quartz, Si, and/or Mo substrates into a solution of either polydopamine (PDA) or polystyrenesulfonate (PSS) and then in the CIGO-PAH dispersion to fabricate films as thick as 1–2 microns. PSS/CIGO-PAH films were found to be inadequate due to weak adhesion to the Si and Mo substrates, excessive particle diffusion during sulfurization, and mechanical softness ill-suited to further processing. PDA/CIGO-PAH films, in contrast, were more mechanically robust and more tolerant of high temperature processing. After LbL deposition, films were oxidized to remove polymer and sulfurized at high temperature under flowing hydrogen sulfide to convert CIGO to CIGS. Complete film conversion from the oxide to the sulfide is confirmed by X-ray diffraction characterization.

INTRODUCTION

Quaternary chalcogenide semiconductors of structure $CuA_xB_{1-x}Z_2$ (where A, B = In, Ga or Zn, Sn; Z = S or Se; $0 \leq x \leq 1$) are among the leading materials candidates under study as absorber layers for conversion of visible and near infrared solar radiation into electricity in photovoltaic devices. These materials offer several important advantages, including composition tunable band gaps for light absorption matched to the solar spectrum, a large knowledge base of their fundamental properties accumulated over decades of research, and photochemical, chemical, and thermal stability, among others. [1]–[5] However, achieving sufficient power/energy conversion efficiencies (i.e., >20%) using appropriate materials and systems that can be prepared at low cost remain fundamental barriers to photovoltaic commercialization. [1].

With regard to the latter, a key issue in lowering costs is the ability to prepare high quality materials and films in large quantities using processes amenable for high throughput manufacturing. Although vacuum techniques such as sputtering [6], [7] and co-evaporation [8]offer exquisite control over composition and deposition of semiconductor absorbers, they remain largely more costly processing techniques. Consequently, significant efforts are being expended to develop alternative non-vacuum synthetic and film deposition routes based on liquid phase processes compatible with high throughput manufacturing. These include such diverse well-developed technologies as electrodeposition, [7], [9] sol-gel/chalcogenization, [10]reactive solution deposition, [11] interfacial self-assembly, [12] and spincoating, dipcoating, doctor-blading, or ink printing, [5], [13]–[17] alone or in combination with subsequent thermal annealing treatments.

One increasingly popular liquid phase dipcoating technology compatible with high throughput manufacturing is layer-by-layer (LbL) deposition. [18], [19] LbL films are formed via alternate exposure of a substrate to separate aqueous solutions or dispersions containing oppositely multi-charged species. A surface charge reversal occurs during substrate treatment with each species, allowing controlled conformal electrostatic deposition of the oppositely-charged material during the next step. For example, polymer multilayer films are readily prepared using solutions of cationic and anionic polyelectrolytes, with film thickness, structure, and morphology controlled by pH, [20], [21] added salt type, [22] ionic strength, [20] polyelectrolyte molecular weight, [23] and/or temperature [24] during the deposition process. Replacement of one or both polyelectrolyte solutions by appropriately charged nanoparticle dispersions permits fabrication of composite materials. [18], [25]–[30].

With regards to solar energy applications, the LbL technique has been increasingly exploited for device fabrication, *albeit on smaller scales often constrained by the availability of large amounts of the component nanoparticles*. For example, Lee and coworkers [31] have deposited SnO_2nanoparticle/polyallylamine (PAH) multilayers that were sintered to prepare SnO_2 films useful for cascadal energy band gap matching in dye sensitized solar cells (DSSCs). Furthermore, Ruhlmann and coworkers [32] have recently fabricated DSSCs via a LbL approach using polyoxometalate and porphyrin dye components. Zotti and coworkers [33] have prepared photovoltaic cells from LbL multilayers comprising PbSe nanocrystals and polyvinylpyridine, evaluating semiconductor stability and properties using photoelectrochemical and photoconductivity techniques. In similar fashion, Nozik and coworkers [16] have described LbL deposition of Schottky solar cell devices prepared from PbSe nanocrystals and ethanedithiol crosslinkers, with an unsintered device exhibiting 2.1% efficiency. More recently, Srestha and coworkers [34] have reported a photovoltaic device incorporating a multilayer absorber layer prepared via LbL deposition of polyethylenimine (PEI) and polystyrenesulfonate (PSS)/oleylamine-coated copper-indium-gallium selenide (CIGSe) nanoparticles, demonstrating 3.5% efficiency for the non-optimized device.

We describe here a unique approach using copper-indium-gallium oxide (CIGO) nanoparticles, *which are readily produced in large quantities*, in combination with automatable LbL deposition methods for the low cost manufacture of copper-indium-gallium sulfide (CIGS) films. Our method involves gram scale preparations of CIGO nanoparticles using a flame spray pyrolysis (FSP) technique, [35]–[37] their subsequent surface modification via binding of polyallylamine (PAH), and the formation of stable aqueous dispersions prepared from the resulting CIGO-PAH colloids. Composite multilayer films comprising the CIGO-PAH colloids, together with PSS or polydopamine (PDA), are further prepared via an aqueous LbL approach and characterized. Subsequent oxidation to remove the organic components and sulfurization to form CIGS films for use as potential light absorber layers in photovoltaic devices are also characterized and discussed.

EXPERIMENTAL SECTION

Materials

All chemicals were used as received from the indicated sources unless otherwise noted. Deionized water (18.2 MΩ·cm) for all experiments was prepared by passing water through an Elix 5 Milli-Q Plus Ultra-Pure Water System (Millipore Corp.). Nitrogen gas was obtained from in-house liquid N_2 boil-off and passed through a cellulose filter prior to use. Methane and oxygen gas for flame spray pyrolysis (FSP) experiments and hydrogen sulfide gas for the sulfurization experiments were from Airgas Inc. Acetone, methanol, hydrochloric acid, sulfuric acid, glacial acetic acid, sodium hydroxide ($\geq$ 99.99%, Electronics Grade), sodium chloride, dopamine hydrochloride, tris(hydroxymethyl) aminomethane (Tris), polyethylenimine (PEI, 750,000 g·mole^{-1}), polyallylamine hydrochloride (PAH, 70,000 g·mole^{-1}; 15,000 g·mole^{-1}; 8,500–11,000 g·mole^{-1}), and polystyrenesulfonate (PSS,

70,000 g·mole^{-1}) were all ACS Reagent Grade except where otherwise noted from Aldrich Chemical Co. Polyallylamine hydrochloride (PAH, 120,000–200,000 g·mole^{-1}) was from Alfa-Aesar. Anhydrous ethanol was 200 proof from the Warner-Graham Company. N-(2-aminoethyl)-3-aminopropyltrimethoxysilane(EDA, >95%) from Gelest Inc. was purified by vacuum distillation (140–142°C; 14–15 mm Hg). High purity (≥ 99.999%) copper (II) nitrate hydrate, gallium (III) nitrate hydrate, and indium (III) nitrate hydrate from Aldrich Chemical Co. were used for the preparation of copper-indium-gallium oxide (CIGO) in the FSP experiments.

Silicon wafers (100 mm diameter, 500–550 μm thickness, <100> orientation, p-type (B doping), resistivity 6–9 Ω, No. PD7403) from Wacker Siltronic Corp. were cut into 25 mm ×50 mm pieces for use in the experiments. Polished quartz slides (25.4 mm ×50.8 mm ×1 mm) were from Quartz Scientific Inc. Silicon wafers and quartz slides were cleaned per the literature method[38] via successive 30 min immersions in 1:1 v/v HCl/CH$_3$OH and concentrated H$_2$SO$_4$, with copious water rinsing after each treatment. An EDA self-assembled monolayer (SAM) was chemisorbed onto the freshly cleaned Si wafer and quartz substrate surfaces by treatment for 25 min in 1% (v/v) EDA aqueous solution containing 1 mM acetic acid, followed by a triple water rinse, drying in the filtered N$_2$ gas stream, and a 6 min bake at 110°C to complete the chemisorption process. [39].

Molybdenum foil (≥ 99.9%, Aldrich Chemical Co., 50 mm ×50 mm ×1 mm) substrates were cut into 25 mm ×50 mm pieces for use in experiments and then degreased by successive 5 min rinses in acetone, methanol, and water. The substrates were next polished to remove surface oxides and stains using 15 μm alumina grit followed by 3 μm diamond grit. The polished Mo substrates were rinsed successively with water, isopropanol, and acetone, followed by sonication in water for 15 min at 80 W power using a Branson Model 2510 Ultrasonic Cleaner/Water Bath to dislodge any polishing grit adhering to the surface before drying in a filtered N$_2$ gas stream. The Mo substrates exhibited a near mirror finish with very faint visible scratches after processing. EDA-coated Si wafers and quartz substrates and polished Mo substrates were each stored in Fluoroware containers until need for experiments. The stored, polished Mo substrates were soaked 1 h in 1/1 v/v HCl/methanol and rinsed with water immediately before deposition of multilayer films. The EDA-coated quartz slides and Si wafers were removed from storage and used directly for multilayer depositions.

Instrumentation

Sonication reactions of polyelectrolyte and CIGO particles were carried out using a Sonicators & Materials Inc. Vibra-Cell Sonicator equipped with a titanium horn. All pH measurements were made using a Corning Pinnacle Model 530 pH meter. A Sorvall Model RC5B Plus Refrigerated Ultracentrifuge equipped with a Sorvall Model SS-34 rotor was used for all centrifugations. CIGO-PAH pellets formed during centrifugation were re-dispersed in water using a Scientific Industries Inc. Model Vortex-2 Genie vortexer. Freeze drying experiments were performed using a VirTis Inc. benchtop K freeze dryer. Automated deposition of some films was carried out using a StratoSequence VI Robot Dipcoater from nanoStrata Inc.

UV-visible absorbance spectra were acquired using a double beam Varian Cary 5000 spectrophotometer. Film spectra were referenced to an EDA-coated quartz slide baseline. Absorbance spectra of various solutions and the CIGO-PAH dispersion were acquired using the same instrument with 0.10 cm pathlength quartz cells referenced to a water blank baseline. Thermogravimetric measurements of the CIGO and CIGO-PAH particles were made using a TA Instruments Inc. Hires TGA2950 Thermogravimetric Analyzer. CIGO-PAH dispersions were characterized using a NanoSight LM10-HSBF nanoparticle characterization system along with corresponding software Nanosight NTA 2.2 (www.nanosight.com; Nanosight Worthington, OH, USA) for particle concentration determinations and a Brookhaven Instruments ZetaPALS dynamic light scattering (DLS) system equipped with 1 cm path-length cuvettes for particle size measurements.

X-Ray Photoelectron Spectroscopy (XPS) measurements was acquired using a Thermo Scientific K-alpha XPS system equipped with Al k- source at 1486 eV. A sample spot size of 400 μm was used. Scanning electron microscopy (SEM) was performed using a Carl Zeiss SMT Supra 55 electron microscope with a Princeton Gamma Tech EDS detector for compositional analysis of films. A Scintag XDS 2000 diffractometer using Cu-k radiation and Rigaku SmartLab x-ray diffractometer using Cu-K radiation were used to collect x-ray diffraction spectra for phase identification of the as-prepared CIGO particles and CIGS films, respectively. A KLA-Tencor AlphaStep D-120 profilometer was used to measure the thickness of the various as–prepared, oxidized, and sulfurized films. A custom Kurt J. Lesker Octos cluster tool was used for film sputtering and electron beam evaporation in the preparation of the photovoltaic test device. A JEOL JEM-2200FS field emission electron microscope was used to image the as-prepared CIGO particles dropcast as an ethanol suspension onto a SPI 200 mesh holey carbon coated Cu TEM grid.

CIGO Particle Preparation

CIGO nanoparticles of composition $CuIn_xGa_{1-x}O_2$ $(0 \leq x \leq 1)$ were prepared using ethanol stock solutions containing appropriate ratios of the copper, indium, and gallium nitrate precursors. For the $x \cong 0.7$ composition (best matched to the solar spectrum upon conversion to CIGS) of interest here, a stock solution was prepared by dissolving 20.0 g (0.1066 moles, anhydrous) copper (II) nitrate hydrate, 23.28 g (0.0774 moles, anhydrous) indium (III) nitrate hydrate, and 9.16 g (0.0358 moles, anhydrous) gallium (III) nitrate hydrate in 200 mL ethanol. The stock solution was fed through a homemade flame spray apparatus nozzle at flow rates of 5 mL·min^{-1} with the aid of an O_2 dispersion gas/ oxidant under a flow rate of 5 L·min^{-1}. Small pilot flames ignited from flowing 1.5 L·min^{-1} CH_4 and 3 L·min^{-1} O_2 and forming a ring pattern were used as the flame ignition source and as a supporting flame for the larger central flame. The pilot flame ringlet surrounded a central capillary tube that sprayed the precursor solution mixed with oxygen dispersant gas to form precursor droplets that underwent combustion in the large central flame. The CIGO powders were either deposited directly on Mo-coated sodalime glass substrates heated by the flame to 400°C or collected on glass fiber filter paper mounted in a water cooled stainless steel collection chimney. For the case of deposition on the filter media the powders were removed by scraping with a Teflon spatula.

Polyelectrolyte Binding to CIGO

The following general procedure, illustrated for PAH (molecular weight 8,500–11,000 g·mole^{-1}), was used for reaction of all amine polyelectrolytes with the CIGO particles. The procedure for reaction of PSS was identical, with the exception that the PSS solution pH was 6.8 rather than the pH 8.2−8.3 used for polyamines.

A 5 mg PAH·mL^{-1} stock solution was prepared by dissolving 1.250 g PAH in 230 mL 1.00 M NaCl (aq) solution in a beaker with stirring. A freshly prepared 6 M NaOH (aq) solution was added dropwise with stirring until the pH = 8.3±0.1. The solution was transferred to a 250 mL volumetric flask, diluted to the mark with 1.00 M NaCl (aq) solution, and stored in a sealed glass bottle under N$_2$ atmosphere until needed for experiments. The 5 mg PAH ·mL^{-1} 1.00 M NaCl (aq) solution had a pH = 8.2±0.1. Similar stock polyelectrolyte solutions were prepared using PEI and PSS (at pH 6.8) in 1.00 M NaCl (aq) as necessary.

A 125 mg sample of CIGO powder and 80 mL of stock 5 mg PAH·mL^{-1} 1.00 M NaCl (pH 8.2) solution were added to a 150 mL high walled Pyrex beaker (80 mm height ×55 mm diameter) resting on a lab jack and securely clamped in place in a well-ventilated fume hood. The height of liquid in the beaker was 35 mm. The sonicator horn was immersed into the mixture such that the bottom of the horn was fixed 18−20 mm above the bottom of the beaker, raising the liquid height to 40 mm. The horn assembly was securely clamped in place and the beaker was loosely wrapped with an Al foil cone extending 10 cm above the top of the beaker as a splash guard. The sonicator power was set to 480 W and the mixture was sonicated for 30 min (*Caution*: noise hazard- ear protection required). Following sonication, the horn and Al foil cone were quickly removed and the temperature of the gray-black dispersion was measured by thermometer. A temperature of 50−55°C was routinely measured, consistent with heating during sonication that reduced the dispersion volume to 60–65 mL.

The dispersion was allowed to cool to room temperature for 20 min, transferred (equal weights) into two centrifuge tubes, and centrifuged for 20 min at 10,000 rpm and 4°C. The clear, blue-green supernatant containing unreacted excess PAH/1.00 M NaCl (aq) solution was decanted and discarded, leaving a gray-black pellet of PAH-modified CIGO particles. Wash water was added to each centrifuge tube in amounts required to provide equal tube weights, the tubes were resealed, and the CIGO-PAH pellet was re-dispersed using a vortexer. The samples were re-centrifuged for 20 min at 10,000 rpm (4°C) and the clear, colorless wash water was discarded from the pellet to complete the first wash cycle. The CIGO-PAH pellet was subjected to 3 such additional wash cycles.

After the final wash, weight measurements indicated that 45–50 mg CIGO-PAH were lost during processing, leaving 75–80 mg CIGO-PAH in the final pellets. The pellets were combined, re-suspended in 80 mL water to provide a 1 mg CIGO-PAH·mL^{-1} aqueous stock dispersion, and stored in tightly sealed BLUE MAX 50 mL Falcon polypropylene conical tubes until needed for film depositions. Unused 1 mg CIGO-PAH·mL^{-1} aqueous stock dispersions were typically discarded after 6 days and replaced by freshly prepared dispersions for film depositions unless noted otherwise.

CIGO-PAH Particle Concentration Determination

Freshly prepared 1 mg CIGO-PAH·mL^{-1} aqueous dispersion was diluted in a series with water until concentrations in the optimal $2–4\times10^8$ particles·mL^{-1} range observable by the NanoSight LM10-HSBF nanoparticle characterization system were obtained. A 60 sec video of particle Brownian motion was taken at room temperature and analyzed to obtain a particle count using the Nanosight NTA 2.2 software. Brightness and gain were the same for all samples and the detection threshold was automatically adjusted by the software. A blur size of 5 pixel ×5 pixel was chosen for all samples. CIGO-PAH particle concentration in the original sample was then calculated from the dilution factors (1/2000) used and the recorded particle counts.

CIGO-PAH Particle Size and Stability Determination

A 3 mL aliquot of 1 mg CIGO-PAH·mL^{-1} aqueous dispersion contained in a 1.00 cm pathlength cuvette was used for each dispersion particle size and stability determination. Ten DLS measurements were taken and averaged in any time point at 22°C. Two stability studies were performed. For long term stability determination, data was acquired once a day until day 13 for the dispersion. A second stability study was conducted for 1 mg CIGO-PAH·mL^{-1} aqueous dispersion containing

20 mM Tris (pH 8.25) buffer. For this study, 60 μL of 1.00 M Tris pH 8.25 aqueous buffer was mixed with 3 mL of the freshly prepared 1 mg CIGO-PAH·mL^{-1} aqueous dispersion. Data was collected continuously for a period of 2 h, after which measurements were taken every 30 min for a period of 8 hr. All samples were kept at room temperature and ambient pressure during the experiments.

Fabrication of PSS/CIGO-PAH Multilayer Films

Sealable, grooved plastic containers designed for storage of microscope slides were used as sample containers for the quiescent hand dipcoated LbL PSS/CIGO-PAH multilayer depositions. Containers were securely positioned such that substrates were vertically oriented during depositions to minimize the chances of binding aggregates or precipitates formed during treatments. Sufficient quantities of 1 mg CIGO-PAH·mL^{-1} dispersion and 5 mg PSS·mL^{-1} 1.00 M NaCl (aq) solution to cover the substrates (*i.e.*, EDA-coated Si wafers and quartz slides) to a depth of 25–30 mm were placed in separate containers. EDA-coated Si wafers (*i.e.*, Si-EDA) and EDA-coated quartz slides (*i.e.*, Q-EDA) were then placed directly into the PSS solution and the container was sealed. After 30 min, the substrates were removed and rinsed for 60 s each with agitation in water. The substrate rinse was repeated two additional times in fresh water before drying in the filtered N$_2$ gas stream.

Samples were then transferred to container holding the CIGO-PAH dispersion for deposition of the CIGO-PAH nanoparticles. Samples were typically treated for 5–6 h at room temperature. Overnight treatments (*i.e.*, ≥ 12 h) resulted in no further appreciable particle deposition (*i.e.*, <2–3%), as measured by UV absorbance spectroscopy at 225 nm or 285 nm. Following the CIGO-PAH treatment, the samples were washed and dried as described for the PSS treatment. This cycle was repeated to build PSS/CIGO-PAH multilayers on the substrate surface.

For the Mo substrate, a polydopamine (PDA) coating was deposited prior to deposition of CIGO-PAH and PSS. Polymerization of dopamine was initiated by addition of a 100 μL aliquot of 1.00 M Tris pH 8.25 (aq) buffer to 10 mL of a freshly prepared 1 mg dopamine hydrochloride·mL^{-1} (aq) solution. The Mo substrate was treated with this solution for 70 min, then removed, rinsed thoroughly three times in fresh water, and dried in the filtered N_2 gas stream. The PDA-coated Mo substrate was then treated as described below with a 1 mg CIGO-PAH·mL^{-1} 20 mM Tris pH 8.25 dispersion to initiate the multilayer deposition. Alternating treatments with the 5 mg PSS·mL^{-1} 1.00 M NaCl (aq) solution and 1 mg CIGO-PAH·mL^{-1} dispersion were then performed as described above until the multilayer film of desired thickness had been deposited.

For multilayers fabricated using the robot dipcoater, 45–50 mL aliquots of 1 mg CIGO-PAH·mL^{-1} dispersion and 5 mg PSS·mL^{-1} 1.00 M NaCl solution were placed into separate treatment beakers. The robot controls were set to refill each of 6 rinse beakers with 70 mL water after each use. Substrates were loaded onto the sample holder and spun during treatments and rinses at 140±5 rpm. Substrates were treated with PSS solution for 30 min and CIGO-PAH dispersion for 90 min, which was the maximum treatment time allowable. The substrates were rinsed 60 s in each of 3 beakers containing fresh water and dried for 90 s in a stream of filtered N_2 gas following deposition of each PSS or CIGO-PAH layer. The CIGO-PAH dispersion and PSS solution were replaced and the sample and rinse beakers thoroughly rinsed with water and dried after deposition of every 4–6 PSS/CIGO-PAH bilayers. Multilayers fabricated using the robot dipcoater were left standing in air in the closed sample chamber following completion of a deposition cycle until the initiation of the next cycle.

For hand dipcoated multilayers comprising large numbers of PSS/CIGO-PAH bilayers requiring fabrication times of days or weeks, samples bearing partially complete films were stored in PSS solution overnight (i.e., 12–16 h). Because PSS depositions were essentially complete (i.e., 100%) within 30 min under our conditions, no further PSS deposition occurred overnight. The absorbance spectra of PSS/CIGO-PAH multilayers on Q-EDA substrates were periodically recorded as the depositions proceeded to monitor film growth. CIGO-PAH dispersion and PSS solution were replenished with fresh aliquots after deposition of every 4–6 PSS/CIGO-PAH bilayers. Completed PSS/CIGO-PAH films were stored in sealed Fluoroware containers until needed for additional experiments. Sample containers holding the CIGO-PAH dispersion and PSS solutions were cleaned after completion of film depositions by soaking overnight in 6 M HCl (aq) solution, which completely dissolved the CIGO-PAH particles, and water, respectively, followed by copious rinsing with water and drying in the filtered N_2 gas stream prior to re-use.

Fabrication of PDA/CIGO-PAH Multilayer Films

The preparation of the PDA/CIGO-PAH multilayers proceeded in similar fashion to the fabrication of the PSS/CIGO-PAH films with the changes noted below. First, freshly prepared PDA was formed *in situ* during substrate treatment by the polymerization of dopamine in basic aqueous solution for the deposition of each PDA layer. A stock aqueous solution containing 1 mg dopamine HCl·mL^{-1} was prepared and stored sealed in the dark at 2–4°C for up to 3 days until needed. A 10 mL aliquot of the stock dopamine HCl solution was allowed to warm to room temperature during 5–10 min, after which 100 µL of 1.00 M Tris pH 8.25 (aq) buffer was added with mixing to initiate PDA polymerization. The substrates to be treated were immediately placed in the solution and allowed to stand for 45 min, during which time the solution developed a yellow-brown color. The substrates were removed from the PDA solution, which was discarded, and rinsed and dried as described for the PSS/CIGO-PAH multilayers. The container holding the PDA solution was rinsed thoroughly with water and dried before re-use for the next PDA layer deposition using freshly prepared PDA. The container was cleaned after completion of film depositions by immersion in KOH-saturated isopropanol for 30 min, followed by copious rinsing with water and drying in the filtered N$_2$ gas stream prior to re-use.

Second, the 1 mg CIGO-PAH·mL^{-1} aqueous dispersion was made basic by addition of 150 µL 1.00 M Tris pH 8.25 buffer to 7.5 ml of the CIGO-PAH dispersion immediately prior to use for film depositions. The PDA-coated substrates were treated 45 min with this CIGO-PAH/Tris pH 8.25 dispersion. Each CIGO-PAH/20 mM Tris pH 8.25 dispersion aliquot was used for fabrication of 2–3 PDA/CIGO-PAH bilayers before replacement with a fresh dispersion aliquot. Lastly, unlike the PSS/CIGO-PAH multilayers, PDA/CIGO-PAH multilayers terminated with a CIGO-PAH layer were stored dry overnight in Fluoroware containers until depositions resumed the next day.

Film Oxidation, Sulfurization, and Photovoltaic Test Device Fabrication

PSS/CIGO-PAH and PDA/CIGO-PAH multilayer films on Si-EDA, Q-EDA, and Mo substrates were thermally annealed 5 h at 550°C in air to remove organic components such PSS, PAH, and PDA and complete the oxidation of the CIGO particles in the films. The oxidized films were converted to the corresponding CIGS films via sulfurization by thermal treatment at 550°C in a flowing 20 ccm H_2S gas stream for 3 h. After film oxidation and sulfurization, the resulting CIGS absorber films were characterized as described in the text.

A CIGS film, prepared by oxidation and sulfurization of a 46 bilayer PDA/CIGO-PAH film deposited on a Mo substrate, was used to prepare photovoltaic devices. This CIGS film was coated with a CdS buffer layer via chemical bath deposition per the literature method. [40] A ZnO/aluminum-doped ZnO (2%) film was then applied to the CdS layer via RF magnetron sputtering, followed by electron beam evaporation of Ni/Al grids to complete the photovoltaic device. [41] Device areas were mechanically scribed to 0.5 cm^{-2}. Devices were illuminated under a calibrated 100 $mW \cdot cm^{-2}$ AM1.5G solar simulator for current-voltage characterization.

RESULTS AND DISCUSSION

CIGO Particles

FSP is a non-vacuum gas phase process capable of producing high purity oxide nanoparticles on an industrial scale. [35]–[37] Volatile or aerosolized liquid fuel containing nanoparticle precursor materials is sprayed into a flame, where the droplets evaporate and undergo combustion. The species formed are rapidly quenched as they leave the reaction zone and deposit as nanoparticle oxides on a cold collector surface. We have for the first time successfully produced gram scale quantities of copper-indium-gallium oxide (CIGO) nanoparticles, having an average composition $CuIn_xGa_{1-x}O_2$, using the technique from ethanol solution containing metal nitrate precursors. Although nanoparticles having compositions with $0 \leq x \leq 1$ can be prepared, we limit our attention here to nanoparticles having $x \cong 0.7$ (as the absorbance of CIGS films of similar composition are well matched to the solar spectrum).

Figure 1, part A, shows the TEM of the as-prepared CIGO nanoparticles. Loose aggregates comprising 10−75 nm diameter particles are typically observed. XRD results in Figure 1, part B, reveal broadened peaks and peak positions consistent with a mixture of oxides, rather than a single phase for the as-formed CIGO particles. Thermogravimetry of the as-formed CIGO particles results in a *weight gain* of 2% during heating to 600°C, consistent with the presence of mixed oxides and/or incompletely oxidized metal species within the particles. Nevertheless, sulfurization readily transforms this material into the corresponding CIGS species, as confirmed by its characteristic XRD pattern [42] in Figure 1, part B. We estimate a density of $\cong 4.0\pm0.4$ g·cm^{-3} for our as-prepared CIGO particles from a volume displacement measurement using a fixed mass of material. Consequently, the as-formed particles do not form stable aqueous suspensions as required for use in layer-by-layer film fabrication processes. Therefore, as a first step in the use of these particles for film fabrication we adapted a literature method [43] for stabilizing our aqueous CIGO particle dispersions.

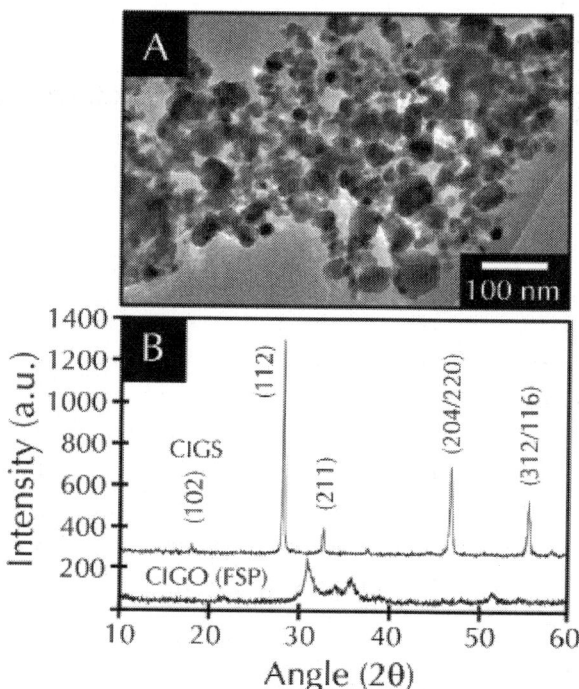

Figure 1: CIGO nanoparticle characterization. (A) TEM of FSP as-prepared CIGO nanoparticles. (B) XRD patterns for the CIGO sample prepared by FSP in part (A) and CIGS species obtained after its sulfurization.

CIGO-PAH Particle Preparation

Amine ligands such as oleylamine are often used as components during thermal syntheses of semiconductor and metal nanoparticles. [44]–[49] Amines readily bind to metal ion surface sites on the growing nanoparticle, controlling nanoparticle size and providing nanoparticle stability during subsequent particle dispersion in solvents. [45], [47], [48] However, because our nanoparticles were prepared via FSP, incorporation of amine ligand during nanoparticle formation is not possible. Therefore, we instead have investigated thermally activated amine ligand binding to our as-formed particles.

We select polyamines, rather than monomeric amines like oleylamine, for further study for three reasons. First, polyamines provide the potential for more rapid surface binding by virtue of their larger numbers of amine sites, with unbound amine sites remaining available for interaction with water to stabilize the particle dispersion. Second, the ability to control surface charge density of the bound polyamine via solution pH changes provides the requisite cationic surface charge necessary for particle use in the electrostatic LbL film assembly process, while also further enhancing particle dispersion stability. Finally, the ability to control thickness of the bound polyamine via changes in polymer molecular weight and solution ionic strength provides a deformable coating expected to facilitate particle packing during film deposition. [25], [26].

Initial work focused on polyethylenimine (PEI), a high molecular weight (750,000 $g \cdot mole^{-1}$) branched polyamine, as the CIGO particle coating. Lvov and coworkers [43] have shown that sonication of colloids in the presence of polyelectrolytes provides the necessary thermal activation for efficient attachment of polyelectrolyte to the colloid particle surface. In addition, sonication agitates and suspends the particles during treatment to promote deposition of a more uniform polyelectrolyte coating.

A sonication treatment was carried out using PEI in 1.00 M NaCl (pH 8.2) solution to promote deposition of thicker films onto the particle surface. Thicker films are deposited from solutions of higher ionic strength because increased counterion pairing with protonated amine sites at higher salt concentrations partially neutralizes polyelectrolyte charge. [22] As repulsive electrostatic interactions among polymer protonated amine sites decrease, van der Waals and related forces become more important leading to polymer chain coiling. Binding of such coiled chains to the particle surface creates a thicker polymer coating on the nanoparticle due to steric repulsions between chains of adjacent polymers, providing additional unbound amine sites for interaction with solvent as required to stabilize particle dispersions. In contrast, extended chains occurring at low ionic strength due to electrostatic repulsions among adjacent protonated amine sites in the polymer produce thinner films with fewer unbound amine sites capable of providing stabilizing interaction with the solvent.

Various experiments under different sonication conditions indicated that 30 min sonication treatment of CIGO particles at 480 W power in aqueous PEI/1.00 M NaCl (pH 8.2) solution was sufficient to bind PEI to the CIGO surface, as evidenced by formation of a stable CIGO-PEI dispersion. Although the temperature of the particle dispersion after sonication was typically 50–55°C, sonication can produce local temperatures ranging from 100s-1000s of degrees which are more than sufficient to promote the amine binding reaction. Cooled dispersions exhibited decreases in liquid volume of 20–25% consistent with heating and appeared gray in color.

During subsequent centrifugation to remove and rinse excess PEI/1.00 M NaCl (pH 8.2) solution from the CIGO-PEI particles, we noted a high degree of particle aggregation indicated by difficulty in re-dispersing the pellet in the wash water. This tendency was further exacerbated if the CIGO-PEI particles were freeze dried prior to re-dispersion. We attributed this behavior to (1) potential non-covalent interactions, such as amine-amine hydrogen bonding and chain entanglements, between extended PEI polymer chains on adjacent particles, and (2) covalent bridging of adjacent nanoparticles by a single surface bound PEI chain. We therefore hypothesized that replacement of branched PEI by a linear polyamine, such as polyallylamine (PAH), of lower molecular weight would minimize aggregation. In fact, re-dispersion of CIGO-PAH particles in water was facilitated relative to CIGO-PEI particles, with particles modified by lower molecular weight PAH generally more readily dispersed consistent with this hypothesis. After a series of experiments, we found that in our hands PAH of molecular weight 8,500−11,000 g·mole^{-1} yielded readily dispersible particles during processing that also remained sufficiently stable as aqueous dispersions for subsequent work (vide infra). Amine binding to the CIGO particle surface was confirmed by a control experiment in which PSS replaced PAH during sonication. CIGO particles treated with PSS rapidly settled following sonication, consistent with the negligible metal ligating ability of the PSS sulfonate group compared to the PAH amine group.

CIGO-PAH Particle Dispersion Characterization

We have characterized the CIGO-PAH particles by several techniques to assess their stability and suitability for LbL deposition. The presence of a PAH coating on the CIGO particles is inferred by their formation of a stable aqueous dispersion and electrostatic binding of the particles to a PSS film deposited onto an EDA-coated quartz slide (*vide infra*). In contrast to the 2% weight gain observed for bare CIGO particles, thermogravimetry indicates a 6% weight *loss* for CIGO-PAH particles consistent with oxidation and loss of the organic coating during a temperature ramp to 550°C.

Measurements of Brownian motion of CIGO-PAH particles dispersed in water, made using a NanoSight LM10-HSBF nanoparticle tracking system, indicate that a 1 mg CIGO-PAH·mL^{-1} dispersion contains $(9\pm2) \times 10^{11}$ particles·mL^{-1} (n = 8), with an average particle size of 125±15 nm and particle distribution shown in Figure 2, part A. While light scattering results measure the total particle diameter comprising the CIGO particle and PAH coating, the 125±15 nm particle size noted for the CIGO-PAH particles compared to the as-prepared 10–75 nm diameter CIGO particles (Figure 1, part A) suggests that some PAH-assisted particle aggregation occurs during sonication despite the use of the lower molecular weight PAH species. The UV-visible absorbance spectrum corresponding to the CIGO-PAH dispersion, shown in Figure 2, part B, is characterized by the rapidly rising absorbance with decreasing wavelengths expected for particle light scattering behavior. The spectrum also features a distinct peak at 285 nm, which is not present in the thermally annealed CIGO particles. Because PAH exhibits no absorbance at wavelengths ≥ 200 nm, we assign this peak to incompletely oxidized species consistent with the weight *gain* observations during thermogravimetric oxidation of the as-prepared CIGO samples (*vide supra*).

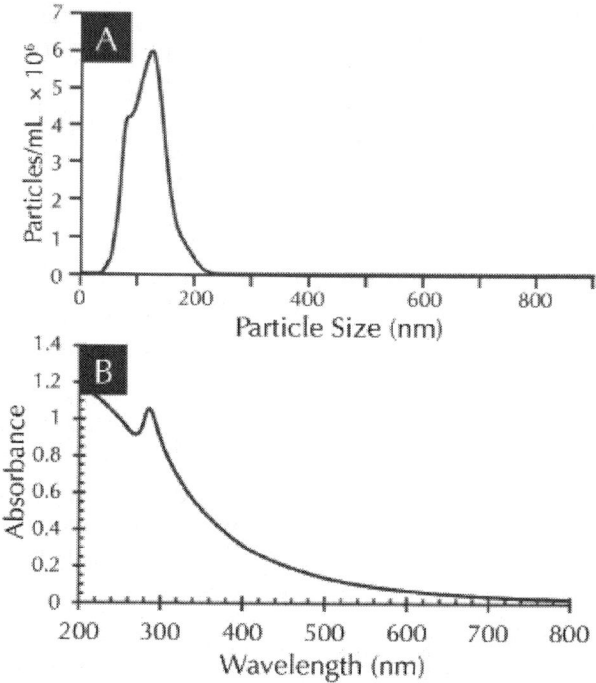

Figure 2: CIGO-PAH dispersion characterization. (A) Particle size distribution and concentration for the fresh 1 mg CIGO-PAH·mL^{-1} (aq) dispersion determined using the NanoSight LM10-HSBF nanoparticle tracking system. (B) UV-visible absorbance spectrum of a freshly prepared 0.33 mg CIGO-PAH·mL^{-1}aqueous dispersion in a b = 0.10 cm pathlength cuvette *vs.* a water baseline. The peak at 285 nm corresponds to incompletely oxidized species within the particles.

Stability of the CIGO-PAH dispersion was also assessed as a prerequisite for its use in the fabrication of PSS/CIGO-PAH and PDA/CIGO-PAH multilayer films. For the CIGO-PAH dispersions used in the preparation of PSS/CIGO-PAH films, changes in particle size distribution over a 13 day period were monitored by dynamic light scattering (DLS). The appearance of particles larger than 300 nm, which were not observed in the freshly prepared dispersion, was set as the aggregation threshold. Aggregates were detected in the dispersion prior to mixing only after 3 days, as shown in Figure 3, part A. However, aggregates were no longer detected upon mixing the sample, indicating that aggregation at this stage was still reversible (Figure 3, part B). In contrast, after 6 days no aggregates were detected in the quiescent sample until the sample was mixed. Aggregates were then detected 10 min after mixing, as shown in Figure 3, part C, and persisted for at least 40 min (Figure 3, part D) indicative of irreversible aggregation. Even larger aggregates were detected at days 7 and 13 after mixing followed by a 10 min wait, consistent with further destabilization of the dispersion as shown in Figure 3, parts E and F. DLS measurements indicated that <5% of the original particles separated from the dispersion as a result of settling during the course of the experiment, a value that did not materially alter particle binding kinetics under our deposition conditions.

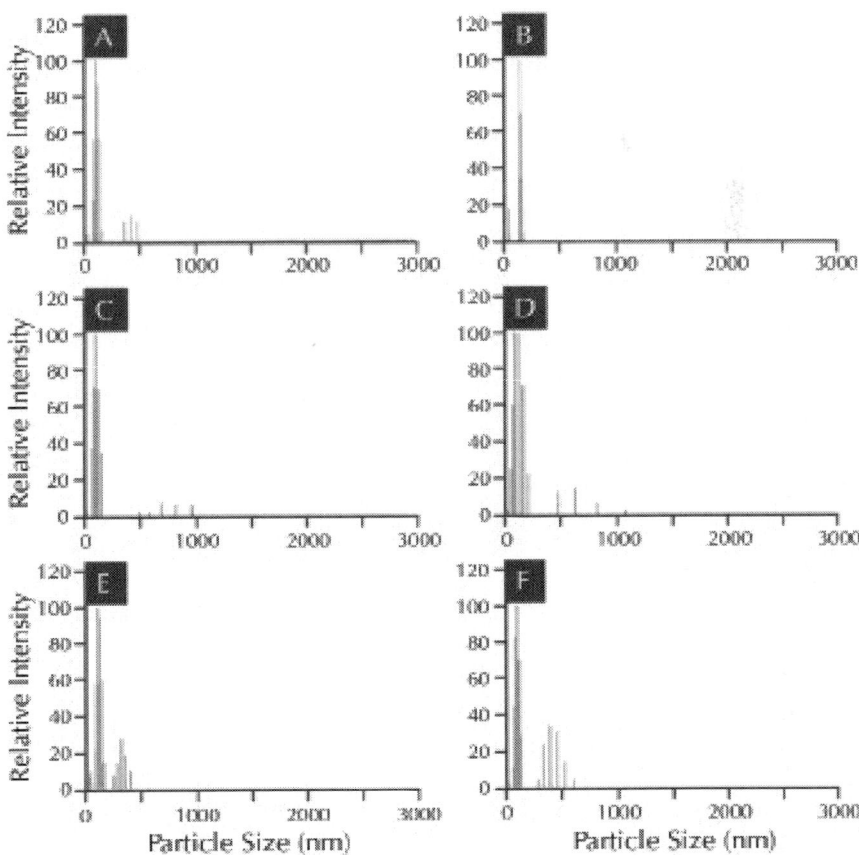

Figure 3: Aqueous CIGO dispersion stability. Particle size dispersion histograms from DLS illustrating the time dependent aggregation of the quiescent

1-PAH·mL⁻¹ (aq) dispersion are shown. (A) aged 3 days; (B) aged 3 days, after mixing; (C) aged 6 days, 10 min after mixing; (D) aged 6 days, 40 min after mixing; (E) aged 7 days, 10 min after mixing; (F) aged 13 days, 10 min after mixing. Particles larger than 300 nm were set as the aggregation threshold.

The stability of freshly prepared aqueous 1 mg CIGO-PAH·mL⁻¹ 20 mM Tris pH 8.25 buffered dispersion used for the deposition of the PDA/CIGO-PAH films was also evaluated, under quiescent conditions, by DLS measurements during an 8 h aging experiment. The average size of the most abundant particles in the dispersion during 8 h was 137±22 nm (n = 22). Some particles larger than 300 nm were first detected by DLS after 30 min and after 50 min some micron size particles were observed, as shown in Figure 4. Aggregation occurred in periodic fashion in which micron size particle levels initially increased to a threshold value, then dropped precipitously as the large aggregates settled out of the dispersion before the cycle began anew. Quantities of micron sized aggregates were detected after 50, 110, 220, and 420 min of aging, with few if any noted at intermediate times consistent with this mechanism. The growth and presence of micron size particles confirmed that Tris pH 8.25 buffer promotes particle aggregation as the dispersion ages, behavior consistent with partial PAH deprotonation ($pK_a \cong 9.5-10$) at pH 8.25 leading to loss of stabilizing positive charge sites on the CIGO-PAH particle. In contrast to the CIGO-PAH dispersion study, a somewhat larger fraction of particles were removed from the CIGO-PAH/Tris pH 8.25 dispersion (<10%) via aggregation and settling. Nevertheless, the CIGO-PAH/Tris pH 8.25 dispersion remained sufficiently stable for use in film fabrication under our deposition conditions.

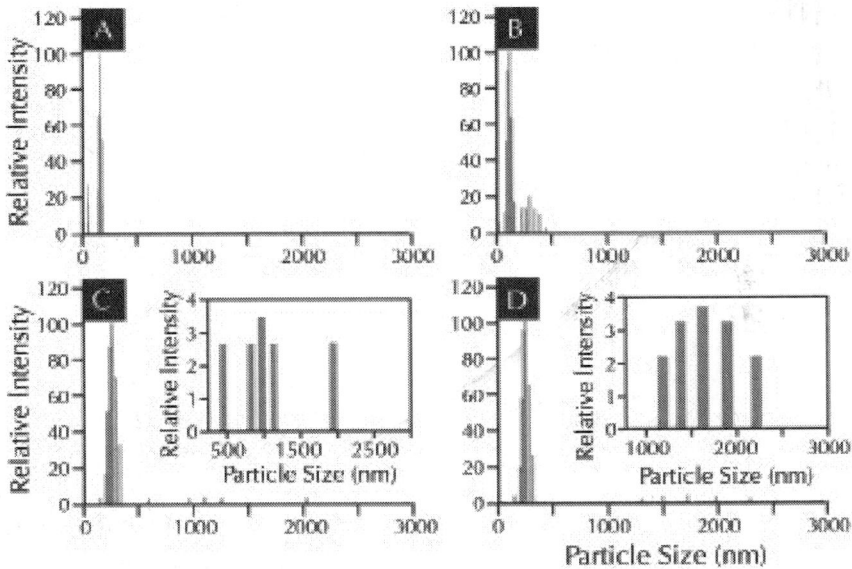

Figure 4: Aqueous CIGO dispersion stability in pH 8.25 buffer. Particle size histograms from DLS illustrating the time dependent aggregation of the 1-PAH·mL^{-1} 20 mM Tris pH 8.25 (aq) dispersion are shown. (A) At preparation (t = 0 min). (B) t = 30 min after preparation. (C) t = 50 min after preparation. (D) t = 420 min after preparation. Insets in parts (C) and (D) are expansions of the particle distributions for particles larger than 400 nm diameter.

PSS/CIGO-PAH and PDA/CIGO-PAH Multilayer Depositions

Figure 5 illustrates the general approach for fabrication, as well as oxidation and sulfurization, of the PSS/CIGO-PAH and PDA/CIGO-PAH multilayers. As a pre-requisite for reproducible fabrication of PSS/CIGO-PAH multilayers, it was first necessary to determine appropriate substrate treatment times for reproducible deposition of CIGO-PAH and PSS layers. For this purpose, a Q-EDA/(PSS/PAH)$_3$/PSS film was prepared using the 5 mg PAH·mL^{-1} 1.00 M NaCl (aq) and 5 mg PSS·mL^{-1} 1.00 M NaCl (aq) solutions as described previously. [22] Treatment of this base film with the CIGO-PAH dispersion indicated that CIGO-PAH particle deposition, as measured by 285 nm film absorbance, was >96% complete after 4 h and >98% complete after 6 h compared to a film deposited overnight (*i.e.*, ≥ 12 h). Similar adsorption studies involving treatment of the CIGO-PAH layer with the 5 mg PSS·mL^{-1} 1.00 M NaCl (aq) solution indicated that PSS binding was complete (*i.e.*, 100%) within 30 min. Consequently, substrate treatment times of 30 min and ≥ 5 h were selected for hand dipcoating depositions of the PSS and CIGO-PAH layers, respectively.

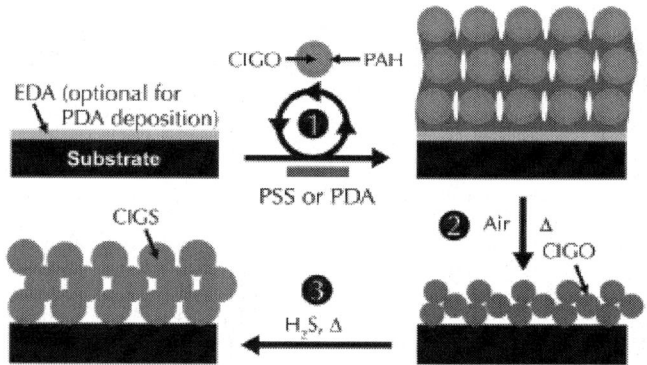

Figure 5: Film fabrication scheme using CIGO-PAH colloids and PSS or PDA polyelectrolytes. Process sequence (not to scale): (1) Treatment of substrate with PSS (aq) or pH 8.25 dopamine (aq) solution (for *in situ* PDA generation) followed by CIGO-PAH (aq) dispersion or CIGO-PAH/Tris pH 8.25 (aq) dispersion to deposit first bilayer of PSS/CIGO-PAH or PDA/CIGO-PAH, respectively. Repetition of treatment cycle deposits additional bilayers for multilayer film fabrication; (2) Air oxidation for 5 h at 550°C to remove organic components from CIGO particles; (3) Sulfurization for 3 h at 550°C in H$_2$S to convert CIGO particles to CIGS film. Consult the Experimental Section and text for additional details.

Once treatment times had been established, PSS/CIGO-PAH multilayer depositions were continued using the PSS/PAH base film to confirm layer deposition reproducibility and assess the effects of CIGO-PAH dispersion age on film growth. Film fabrication was initiated under quiescent hand dipcoating conditions using a freshly prepared stock CIGO-PAH dispersion, which aged 9 days during the time required to complete the film. The CIGO-PAH treatment dispersion was replaced by decantation with the aging CIGO-PAH stock dispersion after every 3 days of use. Care was taken not to transfer any precipitated material present into the treatment container, which was cleaned before the dispersion was replaced. Fabrication of a second film was also separately initiated after the CIGO-PAH stock dispersion was aged 2 days, with the final layers deposited using a CIGO-PAH dispersion aged 11 days.

Figure 6, part A, shows the absorbance spectrum of a completed 18 bilayer PSS/CIGO-PAH multilayer of structure Q-EDA/(PSS/PAH)$_3$/(PSS/CIGO-PAH)$_{18}$, together with spectra of intermediate films having 6 and 12 bilayers. Films having these structures are present on *each* side of the quartz substrate. The spectra are similar to that of the aqueous CIGO-PAH dispersion shown in Figure 2, part A, with the exception of an additional peak near 225 nm characteristic of the PSS layers. The absorbance spectrum of the 18 bilayer film after thermal annealing is also shown. The UV intensity is decreased and the PSS absorbance at 225 nm is absent after annealing due to the removal of the organic PAH and PSS components. In addition, the 285 nm peak, assigned to incompletely oxidized species within the CIGO particles, is also absent, consistent with the thermogravimetry results (*vide supra*).

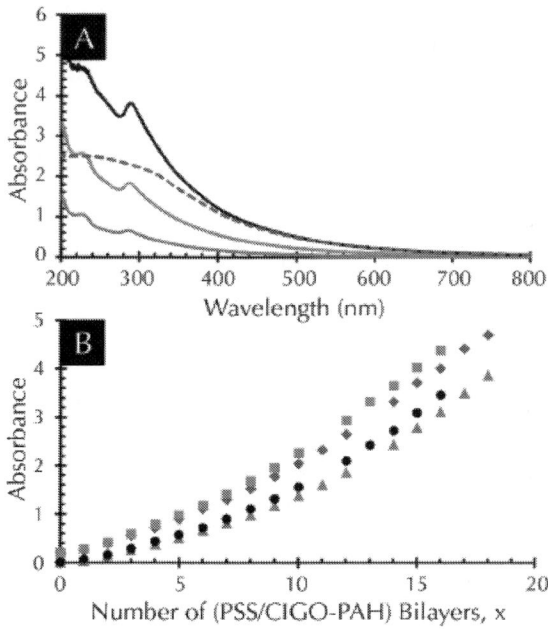

Figure 6: Characterization of PSS/CIGO-PAH multilayers prepared by hand dipcoating. (A) Absorbance spectra in descending order at 285 nm of PSS/

CIGO-PAH multilayer films of structure Q-EDA/(PSS/PAH)$_3$/(PSS/CIGO-PAH)$_x$ with $x = 18$ (black, solid), $x = 18$ after annealing in air 5 h at 550°C (blue, dashed), $x = 12$ (red, solid), and $x = 6$ (green, solid). Note that the measured absorbance represents films having the structures shown that are present on *both sides* of the quartz slide. (B) Absorbance *vs.* number of bilayers, x, for Q-EDA/(PSS/PAH)$_3$/(PSS/CIGOPAH)$_x$ multilayers. Red squares (225 nm) and black circles (285 nm) indicate a film initiated using 2 day aged CIGO-PAH; Blue diamonds (225 nm) and green triangles (285 nm) indicate a film initiated using fresh CIGO-PAH.

Figure 6, part B, shows the changes in PSS/CIGO-PAH film absorbance, monitored at 285 nm where CIGO is the predominant absorbing species and at 225 nm where both CIGO and PSS absorb well, as a function of the number of bilayers deposited for both films. In each case, absorbance increases nonlinearly with slight upwards curvature as the number of bilayers deposited increases. The amount of material deposited is slightly but consistently larger for the film initiated using the 2-day old CIGO-PAH than the fresh CIGO-PAH dispersion. In addition, traces of settled CIGO-PAH precipitate are observed at the bottom of the CIGO-PAH containers following completion of the films, behaviors in agreement with the aggregation noted for the aqueous CIGO-PAH dispersion in the DLS studies. Our results suggest the use of CIGO-PAH dispersions aged at most 6 days, and preferably 3 days or less, to optimize linearity and reproducibility during PSS/CIGO-PAH film depositions again consistent with the DLS results.

Experiments were also performed using a robot dipcoater to demonstrate the ability to automate film fabrication directly onto our substrates for use in a manufacturing environment. Because software constraints associated with the robot limited deposition times to no more than 90 min per layer, substrate treatments were carried out with stirring to enhance deposition rates. Through some trial and error, a 90 min CIGO-PAH deposition with sample stirring ($\sim$140±5 rpm) was found to correspond to a 6 h quiescent CIGO-PAH treatment, as measured by the 285 nm absorbance of the deposited CIGO-PAH material. However, initial attempts to directly deposit multilayers bearing more than 8 PSS/CIGO-PAH bilayers led to complete settling of the CIGO-PAH dispersion. The problem was eventually traced to cumulative carryover of small amounts PSS, identified by its characteristic UV spectrum and 225 nm absorbance band into the CIGO-PAH dispersion. This apparently occurred due to incomplete draining of rinse water during rinse water refill cycles.

Acceptable films could nevertheless be deposited if the CIGO-PAH dispersion and PSS solution were replaced after deposition of every 4–6 bilayers, provided that the sample and rinse beakers were adequately rinsed and dried before re-use. If the used CIGO-PAH dispersion was replaced with dispersion aged 3 days or less, in which aggregate levels are negligible as measured by DLS, further improvements in film quality and reproducibility accrued. Specifically, nearly linear increases in film absorbance as a function of the number of bilayers deposited were now obtained for films comprising as many as 80 bilayers (on *each* side of the quartz slide), as shown in Figure 7, part A. Subsequent oxidation and sulfurization of the film provided an absorber layer capable of complete absorption of visible and near IR light (*i.e.*, absorbance >2 for all wavelengths ≤ 1100 nm), as shown in Figure 7, part B.

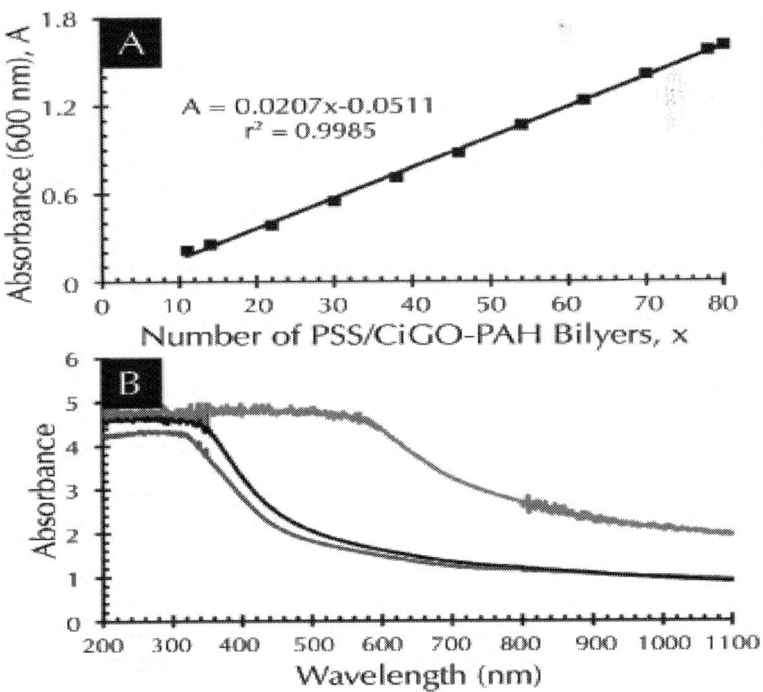

Figure 7: Characterization of PSS/CIGO-PAH multilayers prepared by robot dipcoating. (A) Absorbance *vs.* number of bilayers, *x*, for Q-EDA/(PSS/CIGO-PAH)$_x$ multilayers prepared via automated dipcoating using the robot. (B) Ab-

sorbance spectra in ascending order at 400 nm of robot dipcoated PSS/CIGO-PAH multilayer films of structure Q-EDA/(PSS/CIGO-PAH)$_{80}$ after annealing in air 5 h at 550°C (blue), as deposited (black), and after H$_2$S sulfurization 5 h at 550°C (red). Note that the measured absorbance derives from films having the structures shown that are present on *both sides* of the quartz slide.

Attempts to prepare similar absorbers on a Mo metal substrate useful as a back electrode in photovoltaic devices revealed an additional concern. Although PSS/CIGO-PAH films were successfully deposited, coating migration onto areas of Mo not originally covered by the film was noted following sulfurization. This behavior was *not* observed for a corresponding film deposited on a quartz substrate. It suggested to us that adhesion of the PSS/CIGO-PAH multilayer to the underlying Mo substrate may be insufficient to limit film migration. In fact, an XPS study indicated that neither organothiol and organosiloxane monolayers nor polyamines were readily chemisorbed to the Mo surface (though PSS was weakly bound), behavior consistent with this hypothesis.

Therefore, we sought a polymeric replacement for PSS that could more strongly bind CIGO-PAH particles to each other and the substrate in our multilayer films. PDA is a strong adhesive[50], [51] produced by polymerization of dopamine under basic pH conditions and used by marine organisms such as mussels to anchor themselves to solid surfaces in underwater environments. [52], [53] Although its composition and structure remain uncertain, electrostatic and covalent interactions involving both the amine and catechol sites of dopamine have been identified within PDA. [54]–[56] Therefore, we expected that amine sites on our CIGO-PAH particles would be effectively bound in films containing a PDA component.

In fact, initial experiments revealed deposition of a yellow-brown PDA film onto Q-EDA, Si-EDA, and even Mo substrates treated for extended periods with an aqueous 1 mg dopamine·mL^{-1}solution containing 10 mM Tris pH 8.25 buffer, as shown in Figure 8, part A. A nearly linear deposition rate during the first 45 min was observed in Figure 8, part B, as PDA polymerization occurred on both sides of the EDA-coated quartz substrate. Thereafter, deposition continued at a slower rate (50–60%) suggesting that coverage of the EDA was complete and further reaction occurred on the already deposited PDA layer. However, attempts to bind our CIGO-PAH dispersion directly onto PDA films deposited for 45 min or longer led to irreproducible results. CIGO-PAH was reproducibly deposited only when the dispersion contained 20 mM Tris pH 8.25 buffer, indicating that the presence of catecholate anions on the PDA surface was a pre-requisite for CIGO-PAH binding. Subsequent time dependent studies indicated that CIGO-PAH particle deposition onto a PDA film was complete in as little as 45 min, a substantially shorter time than the 5 h CIGO-PAH treatments required for deposition of the corresponding PSS/CIGO-PAH films in quiescent hand dipcoating experiments.

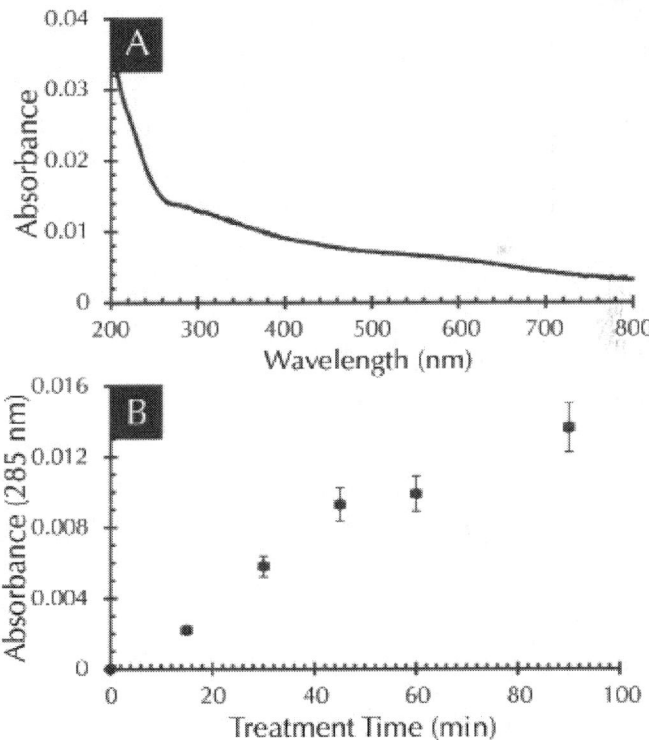

Figure 8: Deposition of polydopamine thin films. (A) Absorbance spectrum of polydopamine (PDA) film on a Q-EDA slide prepared by 90 min treatment in freshly made 1 mg dopamine·mL⁻¹ 10 mM Tris pH 8.25 (aq) solution. Spectrum shown after subtraction of an untreated Q-EDA slide baseline. Note that the measured absorbance represents PDA films that are present on *both sides* of the quartz slide. (B) Time dependent absorbance at 285 nm for Q-EDA slide treated with 1 mg dopamine·mL⁻¹ 10 mM Tris pH 8.25 (aq) buffer solution.

Fabrication of PDA/CIGO-PAH films directly on our substrates via quiescent hand dipcoating was successfully performed using 45 min treatments for both the PDA/Tris pH 8.25 solution and the CIGO-PAH/20 mM Tris pH 8.25 dispersion. By freshly preparing 1 mg dopamine·mL^{-1} 10 mM Tris pH 8.25 solution for deposition of each PDA layer and using a new aliquot of CIGO-PAH/20 mM Tris pH 8.25 dispersion, prepared by addition of 1.00 M Tris pH 8.25 buffer to 1 mg CIGO-PAH·mL^{-1} dispersion aged ≤ 3 days, after deposition of every 3−4 PDA/CIGO-PAH bilayers, a nearly linear and reproducible multilayer deposition was achieved per Figure 9, part A. An absorbance spectrum for the PDA/CIGO-PAH film of structure Q-EDA/(PDA/CIGO-PAH)$_{20}$ present on *each* side of the quartz slide is shown in Figure 9, part B. Spectra are also shown in Figure 9, part B, for the corresponding oxidized and sulfurized films, which are similar to the analogous films prepared from PSS/CIGO-PAH multilayers in Figure 7, part B.

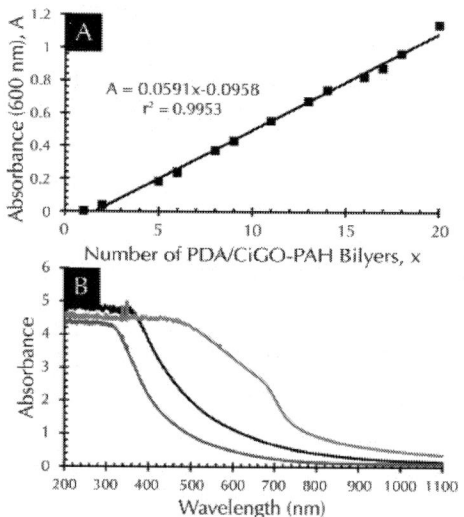

Figure 9: Characterization of PDA/CIGO-PAH multilayers prepared by hand dipcoating. (A) Absorbance *vs.* number of bilayers, *x*, for Q-EDA/(PDA/CIGO-PAH)$_x$ multilayers deposited by hand dipcoating. (B) Absorbance spectra in ascending order at 600 nm of hand dipcoated PDA/CIGO-PAH multilayer films of structure Q-EDA/(PDA/CIGO-PAH)$_{20}$ after annealing in air 5 h at 550°C (blue) with thickness 953±162 nm, as deposited (black) with thickness 1163±189 nm, and after sulfurization 5 h at 550°C (red) with thickness

1079±168 nm. Film thicknesses were measured by profilometry. Note that the measured absorbance derives from films having the structures shown that are present on *both sides* of the quartz slide.

Somewhat thicker films of structure Si-EDA/(PDA/CIGO-PAH)$_{26}$/PDA were also deposited via hand dipcoating on Si-EDA wafers for characterization of film morphology and topography. Top-view and side-view SEM images of the Si-EDA/(PDA/CIGO-PAH)$_{26}$/PDA film as deposited and after oxidation and sulfurization are shown in Figure 10. The as-deposited film in Figure 10, part A, reveals a sponge-like morphology comprising aggregates of ca. 125 nm diameter CIGO-PAH particles cemented together by PDA. The side-view of the film shown in Figure 10, part B, indicates that nanochannels having diameters comparable to the CIGO-PAH particle sizes are present and completely penetrate the film, consistent with inefficient packing of the particles during the deposition process. Film thickness is ca. 1500 nm, with a roughness of ±150 nm again consistent with the presence of pore structures and inefficient particle packing. Although film thickness and roughness (*i.e.*, ca. 1300±100 nm) are each reduced somewhat following air oxidation to remove the PDA and PAH components (Figure 10, parts C and D), porosity, particle size, and general film morphology are little changed.

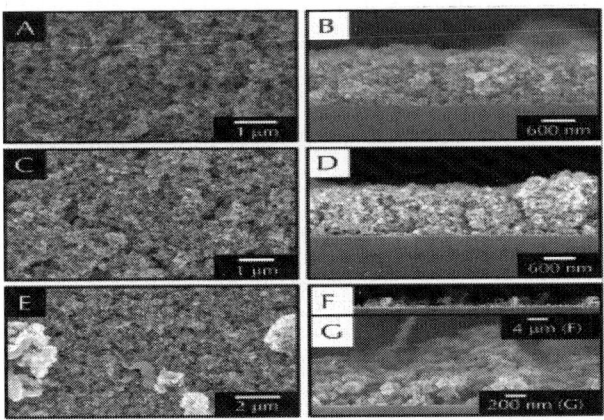

Figure 10: SEM images of Si-EDA/(PDA/CIGO-PAH)$_{26}$/PDA films. Top-views (A, C, E) and side-views (B, D, F, G) for as-deposited film (A, B), as-deposited film after 5 h air oxidation at 550°C (C, D), and oxidized film after sulfurization 3 h in H$_2$S at 550°C (E, F, G) are shown.

However, significant changes are observed following sulfurization. Lighter regions of coalesced material are clearly observed, together with ca. 100 nm particles and pores, on the top surface of the film following sulfurization in Figure 10, part E. The side-view of the same film in Figure 10, part F, shows that the lighter material comprises coiled and rod-shaped fibrils several microns in height distributed over the top surface of the film. A close-up of the film side-view inFigure 10, part G, indicates that the film has thinned significantly, with film thickness in regions absent the fibrils now just ca. 600±100 nm and nanochannels penetrating the film to the underlying Si substrate clearly seen. These observations are collectively consistent with a redistribution of material during the sulfurization process, as also noted during sulfurization of the PSS/CIGO-PAH films deposited on Mo substrates (*vide supra*). The phenomenon is clearly associated with the presence of the substrate and/or polyelectrolytes, since as-prepared CIGO particles are cleanly sulfurized to CIGS (Figure 1, part B) without the dramatic morphology changes observed here. However, the mechanism associated with the film transformation during sulfurization is currently unknown and remains under investigation.

The composition of the sulfurized film shown in Figures 10, parts E–G, is verified by the XRD analysis in Figure 11, which exhibits the reflection peaks expected for CIGS material. No evidence for additional phases or materials is observed, indicating that the fibrils and top surface particulates comprise the same material. This is further confirmed by EDS analyses performed on both the particulate and fibrillar regions. The former exhibits an elemental composition (in atom %) of Cu = 18.35%, In = 21.05%, Ga = 5.07%, and S = 45.26%. A 10.27% oxygen value is also observed, consistent with sampling of the underlying substrate oxide through the film nanochannels during the measurement. EDS of the fibrillar regions provides a substantially identical composition of Cu = 18.73%, In = 20.87%, Ga = 5.50%, and S = 47.10% (with O = 7.81%). Compared to the target CIGS composition of Cu = 25%, In = 17.5%, Ga = 7.5%, and S = 50% expected from our CIGO precursor of composition $CuIn_{0.7}Ga_{0.3}O_2$, our CIGS material is indium rich but both Cu and Ga poor. This result is consistent with our previous observation of a blue-green PAH supernatant characteristic of mixed chloro- and amine- complexes of Cu(II) [57] following CIGO particle sonication in PAH, suggesting that the as-prepared CIGO particle composition may need to be adjusted during FSP to account for selective metal ion extraction during polyelectrolyte binding.

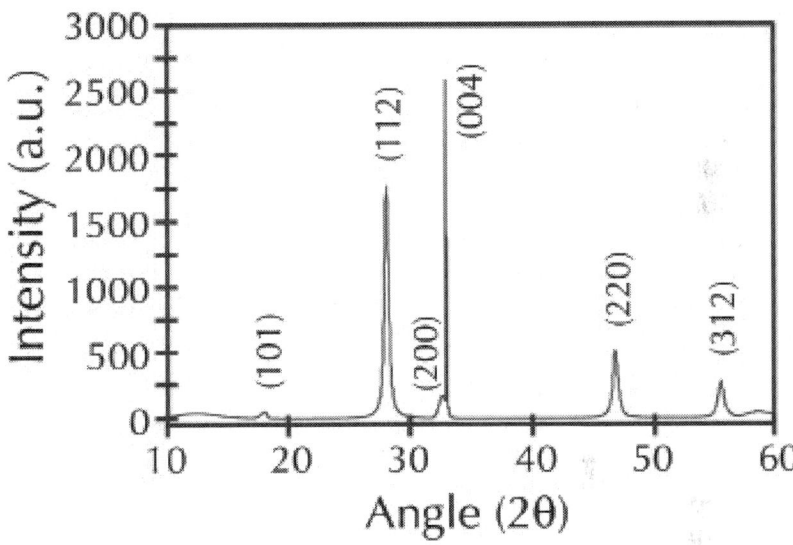

Figure 11: XRD of the CIGS film shown in Figures 9E–G. Consult the text for additional details and discussion.

Finally, we note that attempts to utilize CIGS films similar to those shown in Figure 10, parts E–G, as light absorber components in photovoltaic devices fail to produce a working device, despite the use of thicker precursor PDA/CIGO-PAH films to ameliorate the effects of film thinning and the presence of nanochannels noted in Figure 10, part G. Specifically, a photovoltaic device fabricated from a 46 bilayer PDA/CIGO-PAH film on Mo as described in the Experimental Section exhibits a short circuit response. Such behavior indicates that the drastic morphology changes that accompany sulfurization of our oxidized polyelectrolyte/CIGO-PAH multilayer films, as described and prepared here, remain an impediment to their successful use as light absorber layers in photovoltaic devices.

CONCLUSIONS

We have described procedures for the FSP preparation of gram scale quantities of CIGO particles having diameters of 10−75 nm and tunable compositions from simple solution precursors. Subsequent binding of polyallylamine (PAH) to the particle surface via sonication is accompanied by limited extraction of Cu and Ga from the particle surface and particle aggregation, with the resulting ca. 125 nm diameter CIGO-PAH species forming aqueous colloidal dispersions whose stabilities depend on dispersion age and pH. Dispersions are sufficiently stable, however, for reproducible fabrication of composite PSS/CIGO-PAH multilayer films with PSS via a layer-by-layer hand dipcoating approach. Automation of the process using a robot dipcoater has been demonstrated, providing films containing as many as 80 PSS/CIGO-PAH absorbing >90% of all light of λ ≤ 1100 nm after oxidation and sulfurization. The analogous PDA/CIGO-PAH multilayers are also readily prepared from the CIGO-PAH dispersions via hand-dipcoating using PDA, generated *in situ* by dopamine polymerization, rather than PSS.

Both the PSS/CIGO-PAH and PDA/CIGO-PAH multilayer films are readily oxidized in air to corresponding CIGO films and sulfurized to form CIGS films. The oxidized films retain morphology similar to that of the original multilayers, characterized by fused ca. 125 nm diameter particles and tortuous nanochannels indicative of inefficient particle packing during film deposition. However, drastic morphological changes during film sulfurization lead to the appearance of micron-scale surface fibrils and corresponding thinning of the CIGS film, exposing the underlying substrate via the nanoscale channels present in the film and rendering such films unsuitable for use as light absorber layers in photovoltaic devices.

Optimization of our process to provide more compact CIGS films useful as photovoltaic absorber layers will require addressing both the particle packing issues responsible for film channels and material redistribution phenomenon during sulfurization that promote photovoltaic device short circuit behavior. With regard to packing issues, the use of smaller colloidal particles, produced via jet milling or partial chemical dissolution (*e.g.*, HCl in Experimental Section, *vide supra*) of existing as-deposited CIGO particles (or CIGS particles prepared from them, Figure 1, part B), is expected to provide more uniformly packed films devoid of channels. Film quality may also benefit from enhanced nanoparticle packing via gentle ultrasonic agitation during film deposition [58] and internal restructuring and smoothing of films containing polyelectrolyte components via post-deposition salt annealing. [59].

In the absence of a more detailed understanding of the drastic film morphology changes accompanying our sulfurization process, preparation of LbL films using CIGS particles prepared from the as-deposited CIGO particles are clearly preferred as an alternative fabrication scheme to avoid our CIGO film sulfurization processing step. In fact, preliminary experiments in our laboratory indicate that the binding of polyamines demonstrated for CIGO particles is also applicable for CIGSe particles, as required for fabrication of multilayer films using the LbL approach described here. Additionally, during the course of our work two reports have appeared that further support the viability of this approach. Specifically, Shrestha and coworkers [34] have described a LbL process in which 10 nm diameter CIGSe particles stabilized by bound oleylamine ligands were successfully coated by PSS and, together with PEI, used to fabricate a closely packed PEI/CIGSe-PSS multilayer film. Use of their film as an absorber layer in a photovoltaic device *without* prior thermal annealing to remove the organic components provides a cell exhibiting a solar conversion efficiency of 3.5%. In addition, Korgel and coworkers [17] have shown that thermal annealing under *inert* atmosphere to remove oleylamine adsorbate from CIGSe (*i.e.*, $Cu_{0.8}In_{0.7}Ga_{0.3}Se_2$) nanocrystals in spraycoated nanocrystal films does not materially affect nanocrystal composition and promotes subsequent nanocrystal sintering required for optimization of photovoltaic device performance. We are currently exploring and adapting such techniques in our multilayer systems to fabricate improved quality CIGS films via the LbL approach using our more abundant CIGO particles, or CIGS particles prepared from them, as low cost scalable materials.

ACKNOWLEDGMENTS

We thank Dr. Jas Sanghera, Dr. Ellen Goldman, and Mr. Martin Moore for helpful comments and discussions during the preparation of our manuscript.

AUTHOR CONTRIBUTIONS

Conceived and designed the experiments: WJD CMS CCB JAF WK. Performed the experiments: WJD CMS JF CCB JDM JAF WK. Analyzed the data: WJD CMS JF CCB JDM JAF WK. Contributed reagents/materials/analysis tools: WJD CSM JF CCB JDM JAF. Wrote the paper: WJD WK.

REFERENCES

1. Schmidtke J (2010) Commercial status of thin-film photovoltaic devices and materials. Opt. Express 18: A477–A486. doi: 10.1364/oe.18.00a477

2. Ramasamy K, Malik MA, O'Brien P (2012) Routes to copper zinc tin sulfide Cu_2ZnSnS_4 a potential material for solar cells. J. Chem. Soc. Chem. Commun. 48: 5703–5714. doi: 10.1039/c2cc30792h

3. Hibberd CJ, Chassaing E, Liu W, Mitzi DB, Lincot D, et al. (2010) Non-vacuum methods for formation of $Cu(In, Ga)(Se, S)_2$ thin film photovoltaic absorbers. Prog. Photovoltaics Res. Appl. 18: 434–452. doi: 10.1002/pip.914

4. Wei S-H, Zhang SB, Zunger A (1998) Effects of Ga addition to $CuInSe_2$ on its electronic, structural, and defect properties. Appl. Phys. Lett. 72: 3199–3201. doi: 10.1063/1.121548

5. Panthani MG, Akhavan V, Goodfellow B, Schmidtke JP, Dunn L, et al. (2008) Synthesis of $CuInS_2$, $CuInSe_2$, and $Cu(In_xGa_{1-x})Se_2$ (CIGS) nanocrystal "inks" for printable photovoltaics. J. Am. Chem. Soc. 130: 16770–16777. doi: 10.1021/ja805845q

6. Shi JH, Li ZQ, Zhang DW, Liu QQ, Sun Z, et al. (2011) Fabrication of $Cu(In, Ga)Se_2$ thin films by sputtering from a single quaternary chalcogenide target Prog. Photovoltaics Res. Appl. 19: 160–164. doi: 10.1002/pip.1001

7. Kemell M, Ritala M, Leskelä M (2005) Thin film deposition methods for $CuInSe_2$ solar cells. Crit. Rev. Solid State Mater. Sci. 30: 1–31. doi: 10.1080/10408430590918341

8. Gabor AM, Tuttle JR, Albin DS, Contreras MA, Noufi R, et al. (2004) High-efficiency $CuIn_xGa_{1-x}Se_2$ solar cells made from

$(In_x,Ga_{1-x})_2Se_3$ precursor films. Appl. Phys. Lett. 65: 198–200. doi: 10.1063/1.112670

9. Gougaud C, Rai D, Delbos S, Chassaing E, Lincot D (2013) Electrochemical Studies of One-Step Electrodeposition of Cu-Sn-Zn Layers from Aqueous Electrolytes for Photovoltaic Applications. J. Electrochem. Soc. 160: D485–D494. doi: 10.1149/2.105310jes

10. Kaelin M, Rudmann D, Kurdesau F, Zogg H, Meyer T, et al. (2005) Low-cost CIGS solar cells by paste coating and selenization. Thin Solid Films 480–481: 486–490. doi: 10.1016/j.tsf.2004.11.007

11. Mitzi DB, Yuan M, Liu W, Kellock AJ, Chey SJ, et al. (2009) Hydrazine-based deposition route for device-quality CIGS films. Thin Solid Films 517: 2158–2162. doi: 10.1016/j.tsf.2008.10.079

12. Fontana J, Naciri J, Rendell R, Ratna BR (2013) Macroscopic Self-Assembly and Optical Characterization of Nanoparticle-Ligand Metamaterials. Adv. Opt. Mater. 1: 100–106. doi: 10.1002/adom.201200039

13. Gur I, Fromer NA, Geier ML, Alivisatos AP (2005) Materials science: Air-stable all-inorganic nanocrystal solar cells processed from solution. Science 310: 462–465. doi: 10.1126/science.1117908

14. Wu Y, Wadia C, Ma W, Sadtler B, Alivisatos AP (2008) Synthesis and photovoltaic application of copper(1) sulfide nanocrystals. Nano Lett. 8: 2551–2555. doi: 10.1021/nl801817d

15. Mitzi DB, Yuan M, Liu W, Kellock AJ, Chey SJ, et al. (2008) A high-efficiency solution-deposited thin-film photovoltaic device. Adv. Mater. 20: 3657–3662. doi: 10.1002/adma.200800555

16. Luther JM, Law M, Beard MC, Song Q, Reese MO, et al. (2008) Schottky solar cells based on colloidal nanocrystal films. Nano Lett. 8: 3488–3492. doi: 10.1021/nl802476m

17. Harvey TB, Mori I, Stolle CJ, Bogart TD, Ostrowski DP, et al. (2013) Copper Indium Gallium Selenide (CIGS) Photovoltaic Devices Made Using Multistep Selenization of Nanocrystal Films. ACS Appl. Mater. Interfaces 5: 9134–9140. doi: 10.1021/am4025142

18. Iler RK (1966) Multilayers of Colloidal Particles. J. Coll. Interface Sci. 21: 569–594. doi: 10.1016/0095-8522(66)90018-3

19. Decher G (1997) Fuzzy Nanoassemblies: Toward Layered Polymeric Multicomposites. Science 277: 1232–1237. doi:

10.1126/science.277.5330.1232

20. Lundin M, Solaqa F, Thormann E, Macakova L, Blomberg E (2011) Layer-by-layer Assemblies of Chitosan and Heparin: Effect of Solution Ionic Strength and pH. Langmuir 27, 7537–7548.

21. Wang XF, Sun JK, Ji JA (2011) pH Modulated Layer-by-layer Assembly as a New Approach to Tunable Formulating of DNA within Multilayer Coating. React. Funct. Polym. 71: 254–260. doi: 10.1016/j.reactfunctpolym.2010.09.007

22. Dressick WJ, Wahl KJ, Bassim ND, Stroud RM, Petrovykh DY (2012) Divalent–Anion Salt Effects in Polyelectrolyte Multilayer Depositions. Langmuir 28: 15831–15843. doi: 10.1021/la3033176

23. Porcel C, Lavalle P, Decher G, Senger B, Voegel J-C, et al. (2007) Influence of the Polyelectrolyte Molecular Weight on Exponentially Growing Multilayer Films in the Linear Regime. Langmuir 23: 1898–1904. doi: 10.1021/la062728k

24. Tan HL, McMurdo MJ, Pan G, Van Patten PG (2003) Temperature Dependence of Polyelectrolyte Multilayer Assembly. Langmuir 19: 9311–9314. doi: 10.1021/la035094f

25. Schmitt J, Decher G, Dressick WJ, Brandow SL, Geer RE, et al. (1997) Metal nanoparticle/polymer superlattice films: Fabrication and control of layer structure. Adv. Mater. 9: 61–65. doi: 10.1002/adma.19970090114

26. Bassim ND, Dressick WJ, Fears KP, Stroud RM, Clark TD, et al. (2012) Layer-by-Layer Assembly of Heterogeneous Modular Nanocomposites. J. Phys. Chem. C 116: 1694–1701. doi: 10.1021/jp207912b

27. Kostelansky CN, Pietron JJ, Chen M-S, Dressick WJ, Swider-Lyons KE, et al. (2006) Triarylphosphine-Stabilized Platinum Nanoparticles in Three-Dimensional Nanostructured Films as Active Electrocatalysts. J. Phys. Chem. B 110: 21487–21496. doi: 10.1021/jp062663u

28. Ariga K, Lvov Y, Ichinose I, Kunitake T (1999) Ultrathin films of inorganic materials (SiO_2 nanoparticle, montmorillonite microplate, and molybdenum oxide) prepared by alternate layer-by-layer assembly with organic polyions. Appl. Clay Sci. 15: 137–152. doi: 10.1016/s0169-1317(99)00012-5

29. Sasaki T, Ebina Y, Tanaka T, Harada M, Watanabe M, et al. (2001) Layer-by-Layer Assembly of Titania Nanosheet/Polycation Composite Films. Chem. Mater. 13: 4661–4667. doi: 10.1021/cm010478h

30. Zhou Y, Cheng M, Zhu X, Zhang Y, An Q, et al. (2013) A facile method for the construction of stable polymer–inorganic nanoparticle composite multilayers. J. Mater. Chem. A 1: 11329–11334. doi: 10.1039/c3ta12699d

31. Kim YJ, Kim KH, Kang P, Kim HJ, Choi YS, et al. (2012) Effect of Layer-by-Layer Assembled SnO_2 Interfacial Layers in Photovoltaic Properties of Dye-Sensitized Solar Cells. Langmuir 28: 10620–10626. doi: 10.1021/la3015699

32. Ahmed I, Farha R, Goldmann M, Ruhlmann L (2013) A molecular photovoltaic system based on Dawson type polyoxometalate and porphyrin formed by layer-by-layer self assembly. J. Chem. Soc. Chem. Commun. 49: 496–498. doi: 10.1039/c2cc37519b

33. Vercelli B, Zotti G, Berlin A, Natali M (2010) Self-Assembled Structures of Semiconductor Nanocrystals and Polymers for Photovoltaics. (3) PbSe Nanocrystal–Polymer LBL Multilayers. Optical, Electrochemical, Photoelectrochemical, and Photoconductive Properties. Chem. Mater. 22: 2001–2009. doi: 10.1021/cm903824e

34. Hemati A, Shrestha S, Agarwal M, Varahramyan K (2012) Layer-by-Layer Nanoassembly of Copper Indium Gallium Selenium Nanoparticle Films for Solar Cell Applications. J. Nanomater. 2012: Article No.512409.

35. Pratsinis SE (1998) Flame Aerosol Synthesis of Ceramic Powders. Prog. Energy Combust. Sci. 24: 197–219. doi: 10.1016/s0360-1285(97)00028-2

36. Laine RM, Baranwal R, Hinklin T, Treadwell D, Sutorik A, et al. (1999) Making Nanosized Oxide Powders From Precursors by Flame Spray Pyrolysis. Key Eng. Mater. 159: 17–24. doi: 10.4028/www.scientific.net/kem.159-160.17

37. Mädler L, Kammler HK, Mueller R, Pratsinis SE (2002) Controlled Synthesis of Nanostructured Particles by Flame Spray Pyrolysis. J. Aerosol Sci. 33: 369–389. doi: 10.1016/s0021-8502(01)00159-8

38. Dressick WJ, Dulcey CS, Brandow SL, Witschi H, Neeley PF (1999) Proximity x-ray lithography of siloxane and polymer films containing benzyl chloride functional groups. J. Vac. Sci. Technol. A 17: 1432–1440. doi: 10.1116/1.581833

39. Dressick WJ, Kondracki LM, Chen MS, Brandow SL, Matijević E, et al. (1996) Characterization of a colloidal Pd(II)-based catalyst dispersion for electroless metal deposition. Coll. Surf. A 108: 101–111. doi: 10.1016/0927-7757(95)03392-0

40. Ramanathan K, Contreras MA, Perkins CL, Asher S, Hasoon FS, et al. (2003) Properties of 19.2% efficiency $ZnO/CdS/CuInGaSe_2$ thin-film solar cells. Prog. Photovoltaics Res. Appl. 11: 225–230. doi: 10.1002/pip.494

41. Contreras MA, Egaas B, Ramanathan K, Hiltner J, Swartzlander A, et al. (1999) Progress toward 20% efficiency in $Cu(In,Ga)$ Se_2 polycrystalline thin-film solar cells. Prog. Photovoltaics Res. Applic. 7: 311–316. doi: 10.1002/(sici)1099-159x(199907/08)7:4<311::aid-pip274>3.3.co;2-7

42. Vasekar PS, Dhere NG (2009) Effect of sodium addition on Cu-deficient $CuIn_{1-x}Ga_xS_2$ thin film solar cells. Sol. Energy Mater. Sol. Cells 93: 69–73. doi: 10.1016/j.solmat.2008.04.013

43. Lvov YM, Pattekari P, Zhang X, Torchilin V (2011) Converting Poorly Soluble Materials into Stable Aqueous Nanocolloids. Langmuir 27: 1212–1217. doi: 10.1021/la1041635

44. Thomson JW, Nagashima K, Macdonald PM, Ozin GA (2011) From Sulfur–Amine Solutions to Metal Sulfide Nanocrystals: Peering into the Oleylamine–Sulfur Black Box. J. Am. Chem. Soc. 133: 5036–5041. doi: 10.1021/ja1109997

45. Dilena E, Xie Y, Brescia R, Prato M, Maserati L, et al. (2013) $CuIn_xGa_{1-x}S_2$ Nanocrystals with Tunable Composition and Band Gap Synthesized via a Phosphine-Free and Scalable Procedure. Chem. Mater. 25: 3180–3187. doi: 10.1021/cm401563u

46. Sun Y (2013) Controlled synthesis of colloidal silver nanoparticles in organic solutions: empirical rules for nucleation engineering. Chem. Soc. Rev. 42: 2497–2511. doi: 10.1039/c2cs35289c

47. Tanaka Y, Saita S, Maenosono S (2008) Influence of surface ligands on saturation magnetization of FePt nanoparticles. Appl. Phys. Lett. 92: Article No.093117.

48. Carenco S, Boissière C, Nicole L, Sanchez C, Le Floch P, et al. (2010) Controlled Design of Size-Tunable Monodisperse Nickel Nanoparticles. Chem. Mater. 22: 1340–1349. doi: 10.1021/cm902007g

49. Peng S, Wang C, Xie J, Sun S (2006) Synthesis and Stabilization of Monodisperse Fe Nanoparticles. J. Amer. Chem. Soc. 128: 10676–10677. doi: 10.1021/ja063969h

50. Schaubroeck D, Van Den Eeckhout E, De Baets J, Dubruel P, Van Vaeck L, et al. (2012) Surface modification of a photo-definable epoxy resin with polydopamine to improve adhesion with electroless deposited copper. J. Adhes. Sci. Technol. 26: 2301–2314. doi: 10.1163/156856111x623104

51. Zhang W, Yang FK, Han Y, Gaikwad R, Leonenko Z, et al. (2013) Surface and Tribological Behaviors of the Bioinspired Polydopamine Thin Films under Dry and Wet Conditions. Biomacromolecules 14: 394–405. doi: 10.1021/bm3015768

52. Bandara N, Zeng HB, Wu JP (2013) Marine mussel adhesion: biochemistry, mechanisms, and biomimetics. J. Adhes. Sci. Technol. 27: 2139–2162. doi: 10.1080/01694243.2012.697703

53. Stewart RJ, Ransom TC, Hlady V (2011) Natural Underwater Adhesives. J. Polym. Sci. Part B 49: 757–771. doi: 10.1002/polb.22256

54. Hong S, Na YS, Choi S, Song IT, Kim WH, et al. (2012) Non-Covalent Self-Assembly and Covalent Polymerization Co-Contribute to Polydopamine Formation. Adv. Funct. Mater. 22: 4711–4717. doi: 10.1002/adfm.201201156

55. Kim HW, McCloskey BD, Choi TH, Lee C, Kim M-J, et al. (2013) Oxygen Concentration Control of Dopamine-Induced High Uniformity Surface Coating Chemistry. ACS Appl. Mater. Interfaces 5: 233–238. doi: 10.1021/am302439g

56. Liebscher J, Mrówczy ski R, Scheidt HA, Filip C, H dade ND, et al. (2013) Structure of Polydopamine: A Never-Ending Story? Langmuir 29: 10539–10548. doi: 10.1021/la4020288

57. Cotton FA, Wilkinson G (1988) Advanced Inorganic Chemistry. New York: John Wiley & Sons Inc.

58. Jiang C, Cheng M, Liu H, Shao L, Zeng X, et al. (2013) Fabricating Transparent Multilayers with UV and Near-IR Double-Blocking

Properties through Layer-by-Layer Assembly. Ind. Eng. Chem. Res. 52: 13393–13400. doi: 10.1021/ie401769h

59. Ghostine RA, Jisr RM, Lehaf A, Schlenoff JB (2013) Roughness and Salt Annealing in a Polyelectrolyte Multilayer. Langmuir 29: 11742–11750. doi: 10.1021/la401632x

Comparative Toxicity of Fumigants and a Phosphine Synergist Using a Novel Containment Chamber for the Safe Generation of Concentrated Phosphine Gas

Nicholas Valmas[1] and Paul R. Ebert[1, 2]

[1]School of Molecular and Microbial Sciences, University of Queensland, St Lucia, Queensland, Australia

[2]School of Integrative Biology, University of Queensland, St Lucia, Queensland, Australia

ABSTRACT

Background

With the phasing out of ozone-depleting substances in accordance with the United Nations Montreal Protocol, phosphine remains as the only economically viable fumigant for widespread use. However the development of high-level resistance in several pest insects threatens the future usage of phosphine; yet research into phosphine resistance mechanisms has been limited due to the potential for human poisoning in enclosed laboratory environments.

Principal Findings

Here we describe a custom-designed chamber for safely containing phosphine gas generated from aluminium phosphide tablets. In an improvement on previous generation systems, this chamber can be completely sealed to control the escape of phosphine. The device has been utilised in a screening program with C. elegans that has identified a phosphine synergist, and quantified the efficacy of a new fumigant against that of phosphine. The phosphine-induced mortality at 20°C has been determined with an LC_{50} of 732 ppm. This result was contrasted with the efficacy of a potential new botanical pesticide dimethyl disulphide, which for a 24 hour exposure at 20°C is 600 times more potent than phosphine (LC_{50} 1.24 ppm). We also found that co-administration of the glutathione depletor diethyl maleate (DEM) with a sublethal dose of phosphine (70 ppm, $<LC_5$), results in a doubling of mortality in C. elegans relative to DEM alone.

Conclusions

The prohibitive danger associated with the generation, containment, and use of phosphine in a laboratory environment has now been substantially reduced by the implementation of our novel gas generation chamber. We have also identified a novel phosphine synergist, the glutathione depletor DEM, suggesting an effective pathway to be targeted in future synergist research; as well as quantifying the efficacy of a potential alternative to phosphine, dimethyl disulphide.

INTRODUCTION

The fumigant phosphine (PH_3) is widely used in stored product protection owing largely to its potency, ease of use, and low cost. With the phasing out of ozone-depleting substances in accordance with the United Nations Montreal Protocol, phosphine has become the only economically viable fumigant for the protection of stored grain. The long-term use of phosphine is now under threat however, due to the development of high-level resistance in pest insects. This scenario has motivated efforts to determine the genetic basis of resistance [1]–[2] and to identify the genes that are responsible [3].

Despite decades of study and use, the precise mechanism of phosphine toxicity has not been determined. Toxicity is reliant on molecular oxygen [4]–[6] and is hypothesised to result from inhibition of the mitochondrial complex IV enzyme cytochrome c oxidase, resulting in production of reactive oxygen species and cellular oxidative stress [7]–[11].

Toxicological studies in mammals have shown that glutathione (GSH) provides important protection against phosphine induced disruptions to cells as GSH levels were found to decrease in rat tissue and human blood following phosphine exposure [12]–[15]. Conversely, addition of GSH to mouse cells partially protected them against phosphine-induced cell death, reactive oxygen species, and DNA damage [16]. The GSH depletor buthionine sulfoximine, which irreversibly inhibits γ-glutamylcysteine synthetase at the first step of GSH synthesis, can further lower the already reduced GSH levels in rats exposed to phosphine [15], although the effect on mortality of co-treatment with phosphine and buthionine sulfoximine was not determined. Furthermore, GSH depletion is reported to have no effect on phosphine induced mortality in insects [17].

A practical necessity to prepare for the potential loss of phosphine is identifying a replacement compound. For instance, the botanical insecticide dimethyl disulphide (DMDS) has been suggested as a grain fumigant and is being actively investigated as a soil disinfestant [18]–[19]. Coincidentally, the mode of action of DMDS has recently been proposed to be similar to that of phosphine, in that DMDS treatment results in inhibition of cytochrome c oxidase [20].

The nematode, *Caenorhabditis elegans* has not been widely used as a model organism for studies into fumigant toxicology, despite being ideal in many regards. As a nematode, it represents the class of organism that is the primary target of soil fumigation. It is easy to rear in the laboratory, with a generation time of three days, and due to tremendous reproductive capacity and small size it facilitates rapid screening of compounds for toxicity as well as accurate quantitative analysis. A simple but important practical matter for research purposes is that simultaneous, quantitative application of gaseous and soluble compounds is straightforward with *C. elegans*, whereas it is comparatively difficult with grain-feeding insects. The availability of phosphine resistant mutants [6] and a comprehensive suite of genomic resources (http://www.wormbase.org/) augment the value of *C. elegans* for phosphine resistance research.

Despite the need for phosphine research and the utility of research tools being developed in basic science laboratories, increasingly stringent occupational health and safety regulations in Australia, and likely elsewhere in the world, restrict the use of standard phosphine generation protocols. Such protocols are deemed unsafe in an academic research setting in which facilities are shared by large numbers of student trainees in laboratories often lacking adequate fail-safe systems in the event of power or equipment failure. The problem is compounded by the very high concentrations of gas required to treat organisms that are becoming increasingly resistant to phosphine exposure.

The following report describes a unique device designed specifically for the safe generation of phosphine gas from metal phosphide tablets within an enclosed laboratory, and also demonstrates the use of the soil nematode *C. elegans* as a tool for screening bioactive chemicals. The toxicity of the potential soil fumigant dimethyl disulphide was tested relative to phosphine, using the *C. elegans* system to obtain dose-response curves to the cytochrome coxidase inhibitor, and also to look at potential synergism between phosphine and glutathione depletors.

METHODS

Nematode Culture

Nematodes were grown at 20°C on NGM agar plates (3 g NaCl; 2.5 g peptone; 20 g agar; in 975 mL deionised water, autoclave then add 1mL of 5mg/L cholesterol in ethanol; 1 mL 1 M $CaCl_2$; 1 mL 1 M $MgSO_4$; 25 mL 1M potassium phosphate pH 6). Media containing 2% agar was used rather than the traditional 1.7%, to reduce the incidence of individuals burrowing into the media. Food was provided as a bacterial slurry of *E. coli* OP50 in deionised water.

Nematode eggs were obtained by treating breeding adults with a freshly prepared bleach solution (0.75 N NaOH; 1.5% NaOCl) for 5 minutes, and then rinsing 3 times with M9 buffer (6 g/L Na_2HPO_4; 3 g/L KH_2PO_4; 5 g/L NaCl; 0.25g/L $MgSO_4$ •$7H_2O$). Eggs were left to hatch overnight on an orbital shaker whilst suspended in M9 buffer. Synchronised populations of nematodes were produced by placing newly hatched larvae on NGM agar plates seeded with OP50, at which point they were deemed to be zero hours old.

Chemical Treatment Conditions

Synchronised populations of nematodes were chemically treated when they were 48 hours old. One hour prior to treatment, nematodes were washed off their culture dishes using deionised water and transferred to 12-well tissue culture plates, which contained 2.5 mL of NGM agar per well and were pre-seeded with 20 µL of a 1:30 dilution of OP50 slurry. The number of nematodes per well was then recorded and the plates were left to dry thoroughly before treatment. A dilution of at least 9 individuals per well was desired, in order to amass around 100 nematodes per plate.

Nematodes were chemically treated for 24 hours in glass desiccators that were sealed gas-tight using a rubber O-ring and clamps. Desiccators possessed screw thread adaptors sealed with silicon septa through which phosphine could be injected. After treatment and recovery time, the number of surviving individuals was determined by flooding the culture dish with M9 buffer and observing how many nematodes were freely moving in the aqueous environment. Exposure to either phosphine or DMDS has a narcotic effect on individuals that inhibits development and leaves them paralysed immediately after treatment, making it difficult to score the number of survivors by a motility assay. Therefore, nematodes were left to recover for up to 48 hours before being scored, the exact time being dependant on how quickly the recovering individuals started to produce progeny, which would complicate counting. In situations where there was a compound added to the agar which may affect the nematodes during the recovery period, it was made sure that the air control plates were counted at the same time as the phosphine plates.

Phosphine Generation

Gaseous phosphine was generated by dissolving aluminium phosphide tablets in a sulphuric acid solution and capturing the evolving gas. This procedure was performed in a chamber designed specifically for phosphine production and was located within a fume cupboard to minimize any risk of the gas escaping. The device is shown in figure 1 and consists of two glass vessels with ground flanges around the open ends which allow them to be secured together with a gas-tight seal using a rubber O-ring and clamps. The upper vessel consists of an inner gas receptacle which collects the trapped phosphine; and an outer compartment containing air which is displaced by the production of phosphine and which can be sealed off in the event that the fume cupboard ceases to function. The bottom vessel contains a reservoir of sulphuric acid solution which acts as both a barrier between the phosphine and external environment; as well as catalysing the generation of phosphine from aluminium phosphide tablets. To generate phosphine, the lower vessel is filled with approximately 1 L of 5% sulphuric acid and then a fragment of a Quickphos aluminium phosphide tablet (Bayer CropScience) was dropped into it. An inverted glass funnel was then positioned over the tablet which would trap and channel the gas through the neck and into the central receptacle of the upper vessel. A rubber O-ring was then positioned on the flanges of the lower vessel and the upper vessel was placed on top such that the central receptacle was directly over the funnel neck; and the O-ring was sandwiched between the flanges of the upper and lower vessels. A screw thread adaptor containing a silicon septum was used to seal the inner receptacle, and eight clamps were used to fasten the flanges of the vessels, and create a gas-tight seal. The air trapped within the central receptacle was then completely removed using a syringe, and the device left with the outer compartment remaining unsealed in a fume cupboard whilst the phosphine was produced.

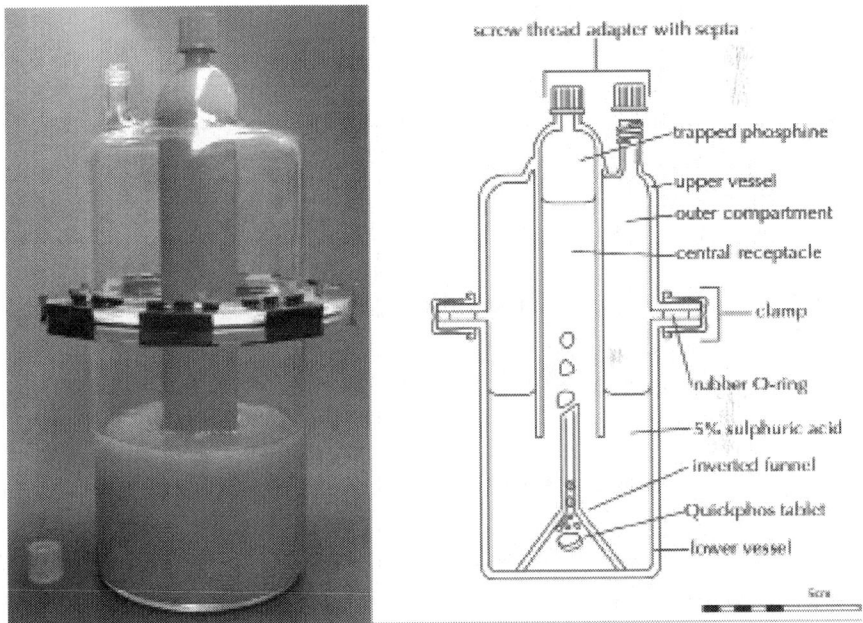

screw thread adapter with septa
trapped phosphine
upper vessel
outer compartment
central receptacle
clamp
rubber O-ring
5% sulphuric acid
inverted funnel
Quickphos tablet
lower vessel

Figure 1: Phosphine Gas Generation Chamber. A photograph of the gas generation chamber is shown (A) with blue liquid representing the sulphuric acid solution; as well as a schematic (B) labelled with the components of the system.

Phosphine Quantification

In addition to aluminium phosphide, Quickphos tablets also contain ammonium carbamate, which prevents phosphine from combusting by decomposing to ammonia and carbon dioxide. As the phosphine generated from Quickphos tablets is not pure, it had to be quantified before use. This was done by injecting a known volume into gas-tight glass desiccator (1 mL into 3 L) and measuring the resulting concentration; which could then be used to determine the concentration of the original gas stock. A continuous flow circuit was established by attaching the glass desiccator to a SmarTox-O gas monitor which had been pre-calibrated on pure phosphine, (The Canary Company Pty Ltd) and recording the phosphine concentration after 5 minutes at which time the mixture was homogenous.

Chemical Administration

Nematode populations were exposed to phosphine by sealing them in gas-tight desiccators and injecting the desired amount of phosphine. Dimethyl disulphide is a volatile liquid at room temperature and was administered by dispensing the required volume onto a glass Petri dish which was sealed inside the desiccator with the nematodes. The volumes of DMDS used in this study evaporate within a few minutes. dl-Buthionine-[S,R]-sulfoximine (BSO) and diethyl maleate (DEM) were added directly to warm NGM agar as the culture plates were being prepared; BSO having been dissolved into deionised water first, while DEM, a liquid at room temperature, was directly added. All NGM agar plates were freshly made within 1 day of being used and were stored in the dark at 4°C to minimised chemical degradation.

Statistical Analysis

Genstat 7 (VSN International Ltd) was used to perform all statistical calculations. Phosphine generation data was analysed using linear regression whilst dose-response data was subject to probit regression. Mortality values for individual biological replicates were adjusted using Abbott's formula, after which the data was pooled for further analysis. Six transformations were performed on the data (probit, logit and complementary log-log on the response variate; and linear and logarithm on the explanatory variate) and regression was performed. The transformation which produced the smallest residual deviance was used to approximate the does-response relationship.

RESULTS AND DISCUSSION

Phosphine Generation

The phosphine generation protocol, which includes the use of a custom designed containment chamber, allows for safe generation of phosphine gas within a laboratory environment. The amount of gas produced by Quickphos tablet fragments of various masses is shown in figure 2, and by using linear regression the relationship between mass and gas volume has been approximated to 232 mL/g. The time taken for a tablet fragment to completely dissolve is dependent on both the mass as well as the shape of the fragment, and due to a lack of consistency in fragment shape it was not possible to establish a relationship between mass and time taken to dissolve. Of all the fragments used in this study however, none took longer than 3½ hours to completely dissolve. The central receptacle of the vessel shown in figure 1a can contain approximately 150 mL of gas, beyond which it will be released into the outer compartment of the upper vessel. Thus, the unit can easily contain the gas produced from 0.65 grams of a Quickphos tablet. Results for chemically pure aluminium phosphide will differ somewhat, because the commercial Quickphos formulation contains not only aluminium phosphide, but also ammonium carbamate.

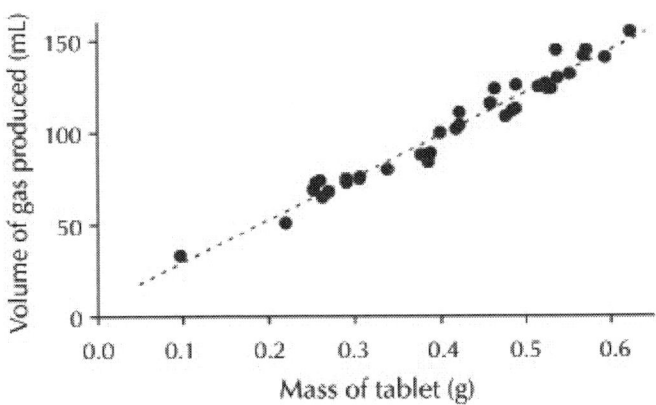

Figure 2: Amount of Gas Produced by Quickphos Fragments. The relationship between the mass of a fragment of Quickphos aluminium phosphide,

and the volume of gas it produces in the chamber shown in figure 1 has been approximated using linear regression. Regression statistics are as follows: slope = 232.48; intercept = 5.99; r^2 = 0.963; n = 34.

The standard phosphine generation protocol uses a device similar to that presented in figure 1 except for the absence of the upper vessel, and consequently the outer chamber which can contain leaking phosphine gas. The standard device did not pass Australian Occupational Health and Safety requirements due to a significant leakage of phosphine from the central gas collection receptacle, through the liquid bath and into the surrounding air. The rate of leakage is low due to the low solubility of phosphine in aqueous solutions, but in the vicinity of the bath, the phosphine concentration reached 1 ppm, which exceeded the permissible exposure limit (PEL) of 0.3 ppm. This was deemed to be an unacceptable risk in the event of a failure of the laboratory ventilation system.

The new chamber includes an upper vessel with a screw thread adaptor that enables the chamber to be completely sealed in the event of a fume cupboard malfunction, thereby preventing any gas escape. We generally allow the gas to be generated completely prior to sealing the septum. In this way, we avoid the pressure build up that otherwise occurs due to gas generation in a restricted volume. It is also possible, however, to carry out the entire phosphine generation process while the chamber is sealed, as there is approximately 800 mL of air above the aqueous bath which acts as a pressure buffer. The chamber was stress-tested to determine its ability to withstand pressure by injecting a large volume of air into the sealed chamber through the septum above the gas collecting chamber. It was found that the chamber could safely contain up to 300 mL of gas. Further addition of gas breached the gas-tight seal by blowing out the O-rings from the flanges of the vessel.

Phosphine and Dimethyl Disulphide Toxicity

The wild-type *C. elegans* line N2 was exposed to a range of phosphine concentrations and the mortality calculated (figure 3). Probit/linear regression estimates that the LC_{50} for a 24 hour fumigation period for this line at 20°C is 732 ppm (95% CI: 708 to 757 ppm). This value is 4 times that previously reported for the LC_{50} of a 24 hour fumigation at 25°C, ~185 ppm (0.26 mg/L) [6]. This is consistent with previous observations of a positive correlation between temperature and toxicity of phosphine in several insect species [21]–[28] and also in rats [29]. This result is explained as an increase in the uptake and metabolism of phosphine by the animals due to higher metabolic rates at higher temperatures [27].

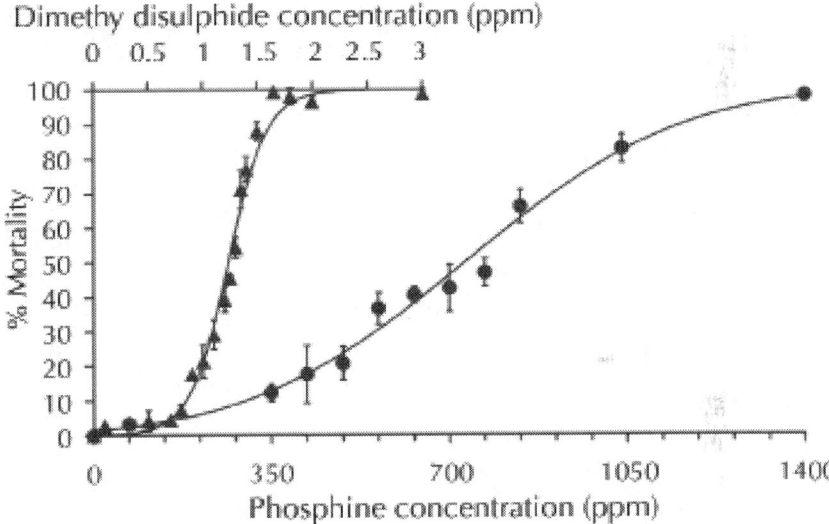

Figure 3: Phosphine and Dimethyl Disulphide Mortality of N2 at 20°C. Mortality of wild-type (N2) *C. elegans* when exposed at 20°C for 24 hours to phosphine (•) and dimethyl disulphide (▲). Regression lines are based on probit/linear and logit/linear relationships respectively. Data points are weighted means from biological replicates±weighted SEM. The LC_{50} of phosphine at 20°C for N2 is 732 ppm; and for DMDS is 1.24 ppm. Plates were counted as follows: 0 ppm phosphine and 0 ppm to 1.1 ppm DMDS, immediately after fumigation; 70 ppm phosphine and 1.2 ppm to 1.5 ppm DMDS, after 24 hours recovery; doses above 70 ppm phosphine and 1.5 ppm DMDS, after 48 hours recovery.

The toxicity of DMDS toward *C. elegans* was determined following a 24 hour exposure at 20°C (figure 3). The LC_{50} is 1.24 ppm (95% CI: 1.20 to 1.27 ppm) as estimated using logit/linear regression. The LC_{50} for DMDS is less than 1/600th that of phosphine. Thus, while the mechanism of action has been proposed to be similar between the two fumigants, DMDS is dramatically more toxic toward *C. elegans* than is phosphine. Whilst no other comparative toxicity studies have been carried out between phosphine and DMDS, independent experiments have been reported for the cowpea weevil *Callosobruchus maculatus* [20],[30]–[33]. Comparison of LC_{50} values for 24 hours treatments with each fumigant supports the general conclusion that DMDS is much more toxic than is phosphine; a fact made inconspicuous because the culture and fumigation conditions between the studies were not identical. It does appear that the difference in toxicity between the two compounds is about an order of magnitude greater in the nematode, *C. elegans* than in the insect, *C. maculatus*. As the physical properties of DMDS make it most suitable as a soil fumigant for which nematodes are the primary target, its toxicity toward nematodes is a prime consideration. The extreme toxicity of this compound as well as the ability of the human nose to detect concentrations of this chemical well below the allowable exposure limits bode well for the efficacy and safety of DMDS as a soil fumigant.

Phosphine and Diethyl Maleate Synergism

A sub-lethal dose of phosphine (70 ppm) was tested together with the GSH depletors DEM and BSO. At the concentrations tested (1 µM, 10 µM, 100 µM, 1 mM, 10 mM) BSO failed to cause any mortality either by itself, or in combination with phosphine (data not shown). Lethal doses of BSO could not be achieved as it was not practical to dissolve BSO in the growth medium at concentrations greater than 10 mM. It is likely that the high tolerance of *C. elegans* to BSO is due to the hydrophilic character of the compound which likely prevents it from penetrating the hydrophobic cuticle of the nematodes. Using the more hydrophobic GSH depletor diethyl maleate, it was possible to induce mortality in *C. elegans* (figure 4). The LC_{50} following 48 hour exposure at 20°C was determined to be 5.98 mM (95% CI: 5.634 to 6.3 mM) by complementary log-log/log regression. A constant, sub-lethal concentration of phosphine (70 ppm) caused a synergistic doubling in mortality due to exposure to DEM, with an LC_{50} of 2.896 mM (95% CI 2.719 to 3.063 mM). This is the first report of phosphine and a glutathione depletor acting synergistically to increase mortality, as a previous study [17] reported no change in phosphine susceptibility in insects treated with BSO.

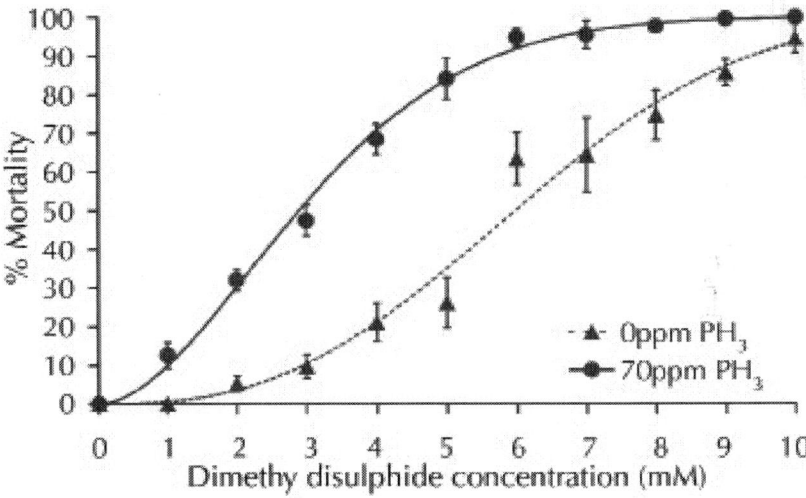

Figure 4: Diethyl Maleate Interaction with Phosphine. Mortality of wild-type (N2) *C. elegans* when exposed to diethyl maleate and two different doses of phosphine: 0 ppm (▲); and 70 ppm (●); at 20°C for 24 hours. Regressions lines are based on complementary log-log/log relationships, and data points are weighted means from biological replicates±weighted SEM. The LC_{50} of DEM in the absence of phosphine at 20°C for N2 is 5.98 mM; and with 70 ppm PH_3 is 2.9 mM. All plates were counted after 24 hours recovery.

CONCLUSIONS

In the present study we describe a unique phosphine generation chamber that allows for the safe production and containment of the gas in a laboratory environment. We use this device and the model organism *C. elegans*, as part of a screening protocol for the assessment of chemical toxicity relative to, and in conjunction with, phosphine. The toxicity of dimethyl disulphide supports its development as a soil fumigant. Co-treatment with phosphine and diethyl maleate identified for the first time, a protective mechanism against phosphine exposure in invertebrates that had previously been observed in mammals. It is hoped that improved handling of the poisonous gas will encourage research on the fumigant, especially with novel research strategies in academic research labs, so that more may be understood about the pathways of toxicity and resistance.

ACKNOWLEDGMENTS

We thank Bob Pryke and Ian Crighton of the UQ Glassblowing Service (University of Queensland) for constructing equipment used in this study, Dan Martin of UQ OH&S for very helpful discussion and Steven Zuryn and Jujiao Kuang for experimental assistance.

AUTHOR CONTRIBUTIONS

Conceived and designed the experiments: NV PE. Performed the experiments: NV. Analyzed the data: NV. Contributed reagents/materials/analysis tools: PE. Wrote the paper: NV PE.

REFERENCES

1. Bengston M, Collins PJ, Daglish GJ, Hallman VL, Kopittke R, et al. (1999) Inheritance of phosphine resistance in *Tribolium castaneum* (Caleoptera : Tenebrionidae). J Econ Entomol 92: 1–20.

2. Collins PJ, Daglish GJ, Bengston M, Lambkin TM, Pavic H (2002) Genetics of resistance to phosphine in *Rhyzopertha dominica* (Coleoptera : Bostrichidae). J Econ Entomol 95: 862–869.

3. Schlipalius DI, Cheng Q, Reilly PEB, Collins PJ, Ebert PR (2002) Genetic linkage analysis of the lesser grain borer *Rhyzopertha dominica* identifies two loci that confer high-level resistance to the fumigant phosphine. Genetics 161: 773–782.

4. Bond EJ, Monor HAU (1967) The role of oxygen on the toxicity of fumigants to insects. J Stored Prod Res 3: 295–310.

5. Kashi KP (1981) Response of five species of stored-product insects to phosphine in oxygen-deficient atmospheres. Pestic Sci 12: 111–115.

6. Cheng Q, Valmas N, Reilly PEB, Collin PJ, Kopittke R, et al. (2003) *Caenorhabditis elegans* mutants resistant to phosphine toxicity show increased longevity and cross-resistance to the synergistic action of oxygen. Toxicol Sci 73: 60–65.

7. Chefurka W, Kashi , KP , Bond , EJ (1976) Effect of phosphine on electron-transport in mitochondria. Pestic Biochem Physiol 6: 65–84.

8. Kashi KP, Chefurka W (1976) Effect of phosphine on absorption and circular dichroic spectra of cytochrome-c and cytochrome-oxidase. Pestic Biochem Physiol 6: 350–362.

9. Jian F, Jayas DS, White NDG (2000) Toxic action of phosphine on the adults of the copra mite *Tyrophagus putrescentiae* [Astigmata : Acaridae]. Phytoprotection 81: 23–28.

10. Dua R, Gill KD (2004) Effect of aluminium phosphide exposure on kinetic properties of cytochrome oxidase and mitochondrial energy metabolism in rat brain. Biochim Biophys Acta 1674: 4–11.

11. Singh S, Bhalla A, Verma SK, Kaur A, Gill K (2006) Cytochrome-*c* oxidase inhibition in 26 aluminium phosphide poisoned patients. Clin Toxicol 44: 155–158.

12. Chugh SN, Kolley T, Kakkar R, Chugh K, Sharma A (1997) A critical evaluation of intravenous magnesium in acute aluminium phosphide poisoning. Magnes Res 10: 225–230.

13. Hsu C-H, Han B-C, Liu M-Y, Yeh C-Y, Casida JE (2000) Phosphine-induced oxidative damage in rats: attenuation by melatonin. Free Radic Biol Med 28: 636–642.

14. Hsu C-H, Chi B-C, Casida JE (2002) Melatonin reduces phosphine-induced lipid and DNA oxidation *in vitro* and *in vivo* in rat brain. J Pineal Res 32: 53–58.

15. Hsu C-H, Chi B-C, Liu M-Y, Li J-H, Chen C-J, et al. (2002) Phosphine-induced oxidative damage in rats: role of glutathione. Toxicology 179: 1–8.

16. Hsu C-H, Quistad GB, Casida JE (1998) Phosphine-induced oxidative stress in Hepa 1c1c7 cells. Toxicol Sci 46: 204–210.

17. Chaudhry MQ, Price NR (1992) Comparison of the oxidant damage induced by phosphine and the uptake and tracheal exchange of ^{32}P-radiolabelled phosphine in the susceptible and resistant strains of *Rhyzopertha dominica* (F.) (Coleoptera: Bostrychidae). Pestic Biochem Physiol 42: 167–179.

18. Coosemans J (2005) Dimethyl disulphide (DMDS): a potential novel nematicide and soil disinfectant. Acta Hortic 698: 57–63.

19. Fritsch J (2005) Dimethyl disulfide as a new chemical potential alternative to methyl bromide in soil disinfestation in France. Acta Hortic 698: 71–75.

20. Dugravot S, Grolleau F, Macherel D, Rochetaing A, Hue B, et al. (2003) Dimethyl disulfide exerts insecticidal neurotoxicity through mitochondrial dysfunction and activation of insect K_{ATP} channels. J Neurophysiol 90: 259–270.

21. Sato K, Higuchi Y, Suwanai M (1973) Studies on the characteristics of action of fumigants I. The 50 percent knock-down dose of hydrogen phosphide to azuki bean weevil *Callosobruchus chinensis* L., calculated from the uptake amounts of oxygen by the weevil. Botyu-Kagaku 38: 22–25.

22. Bell CH (1976) Tolerance of developmental stages of 4 stored product moths to phosphine. J Stored Prod Res 12: 77–86.

23. Hole BD, Bell CH, Mills KA, Goodship G (1976) The toxicity of phosphine to all developmental stages of thirteen species of stored product beetles. J Stored Prod Res 12: 235–244.

24. Bell CH (1977) Toxicity of phosphine to the diapausing stages of *Ephestia elutella, Plodia interpunctella* and Other Lepidoptera. J Stored Prod Res 13: 149–158.

25. Vincent LE, Lindgren DL (1977) Toxicity of methyl-bromide and phosphine at various temperatures and exposure periods to metamorphic stages of *Lasioderma serricorne*(Coleoptera: Anobiidae). J Econ Entomol 70: 497–500.

26. Bond EJ (1984) Manual of Fumigation for Insect Control. FAO Plant Production and Protection Paper No. 54, 1984, FAO, Rome.

27. Chaudhry MQ, Bell HA, Savvidou AD, MacNicoll AD (2004) Effect of low temperatures on the rate of respiration and uptake of phosphine in different life stages of the cigarette beetle *Lasioderma serricorne* (F.). J Stored Prod Res 40: 125–134.

28. Hartsell PL, Muhareb JS, Arnest ML, Hurley JM, McSwigan BJ, et al. (2005) Efficacy of a mixture of phosphine/carbon dioxide on eight species of stored product insects. Southwest Entomol 30: 47–54.

29. Muthu M, Krishnakumari MK, Muralidhara , Majumder SK (1980) A study on the acute inhalation toxicity of phosphine to albino rats. Bull Environ Contam Toxicol 24: 404–410.

30. Ahmed S, Khan MA, Ahmad N (2002) Determination of susceptibility level of phosphine in various strains of dhora (*Callosobruchus maculatus* F.). International Journal of Agriculture & Biology 4: 329–331.

31. Dugravot S, Sanon A, Thibout E, Huignard J (2002) Susceptibility of *Callosobruchus maculatus* (Coleoptera: Bruchidae) and its parasitoid *Dinarmus basalis* (Hymenoptera: Pteromalidae) to sulphur-containing compounds: consequences on biological control. Environ Entomol 31: 550–557.

32. Hasan M, Reichmuth C (2003) Advances in stored product protection. pp. 656–661. Proceedings of the 8th International Working Conference on Stored Product Protection, York, UK, 22–26 July 2002.

33. Dugravot S, Thibout E, Abo-Ghalia A, Huignard J (2004) How a specialist and a non-specialist insect cope with dimethyl disulfide produced by *Allium porrum*. Entomol Exp Appl 113: 173–179.

Chapter 7

Vehicle Exhaust Gas Clearance by Low Temperature Plasma-Driven Nano-Titanium Dioxide Film Prepared by Radiofrequency Magnetron Sputtering

Shuang Yu[1], Yongdong Liang[2], Shujun Sun[1], Kai Zhang[1], Jue Zhang[1,2], and Jing Fang[1,2]

[1]Academy for Advanced Interdisciplinary Studies, Peking University, Beijing, China

[2]College of Engineering, Peking University, Beijing, China

ABSTRACT

A novel plasma-driven catalysis (PDC) reactor with special structure was proposed to remove vehicle exhaust gas. The PDC reactor which consisted of three quartz tubes and two copper electrodes was a coaxial dielectric barrier discharge (DBD) reactor. The inner and outer electrodes firmly surrounded the outer surface of the corresponding dielectric barrier layer in a spiral way, respectively. Nano-titanium dioxide (TiO_2) film prepared by radiofrequency (RF) magnetron sputtering was coated on the outer wall of the middle quartz tube, separating the catalyst from the high voltage electrode. The spiral electrodes were designed to avoid overheating of microdischarges inside the PDC reactor. Continuous operation tests indicated that stable performance without deterioration of catalytic activity could last for more than 25 h. To verify the effectiveness of the PDC reactor, a non-thermal plasma (NTP) reactor was employed, which has the same structure as the PDC reactor but without the catalyst. The real vehicle exhaust gas was introduced into the PDC reactor and NTP reactor, respectively. After the treatment, compared with the result from NTP, the concentration of HC in the vehicle exhaust gas treated by PDC reactor reduced far more obviously while that of NO decreased only a little. Moreover, this result was explained through optical emission spectrum. The O emission lines can be observed between 870 nm and 960 nm for wavelength in PDC reactor. Together with previous studies, it could be hypothesized that O derived from catalytically O_3 destruction by catalyst might make a significant contribution to the much higher HC removal efficiency by PDC reactor. A series of complex chemical reactions caused by the multi-components mixture in real vehicle exhaust reduced NO removal efficiency. A controllable system with a real-time feedback module for the PDC reactor was proposed to further improve the ability of removing real vehicle exhaust gas.

INTRODUCTION

Vehicle exhaust gas contains hydrocarbons, nitrogen oxides, carbon dioxide, carbon monoxide, sulphur dioxide, carbon particles, fine particulate matter and small amounts of aromatic hydrocarbons (benzene) and dioxins. Among these pollutants, hydrocarbons are a major contributor to smog, especially in urban areas. Prolonged exposure to hydrocarbons can cause asthma, liver disease, lung disease and cancer. Carbon monoxide reduces the ability of blood to carry oxygen, and overexposure to carbon monoxide poisoning can be fatal. Nitrogen oxides(NO_x), which is a mixture of NO, N_2O, and NO_2, is generated when nitrogen in the air reacts with oxygen at the high temperature and pressure inside the engine. NO_x is a precursor to smog and acid rain.

In the past few decades, non-thermal plasma (NTP) has been widely used to remove volatile organic compounds (VOC) and NO_x [1]–[2]. However, the application of NTP has been greatly restricted by its low energy efficiency and poor CO_2 selectivity. Besides, undesirable byproducts (such as ozone), need to be further treated. Recently, these problems were solved to some extent by a combination of non-thermal plasma with catalyst, so called plasma-driven catalysis. As a promising technology, this technique integrates the advantages of high selectivity from catalysis and fast ignition from plasma, which maintained high energy efficiency and mineralization rate with low by-product formation [3]–[4].

Plasma-driven catalysis (PDC) is a physical and chemical reaction. The reactive species from plasma, such as ions, electrons, excited atomics, molecules and radicals, generate considerable micro-discharge on the surface of the dielectric. These reactive species, especially high energy electrons, contain a large amount of energy, which will activate nearby catalyst and lower the activation energy of the reaction. Internal transition of high-energy particles will generate ultraviolet radiation [5]–[6]. If absorbed energy is greater than the band gap, the electron inside the semiconductor will be excited with a transition from the valence band to the conduction band, which will form electron-hole pairs and induce a series of further redox reactions. Photo-excited holes have a strong ability to obtain electrons, resulting in reacting with the hydroxide ions together with water adsorbed on the catalyst surface and then generating hydroxyl radicals, which leads to a further oxidation of pollutants. Compared with common catalysts, plasma-driven catalysts have many unique advantages, such as high distribution of reactive species, decreased energy consumption, enhanced catalytic activity and selectivity as well as the reduction of the sensitivity to poison [7]–[8].

Compared with non-thermal plasma, the addition of catalysts could significantly enhance the VOC removal efficiency with increased CO_2 selectivity and carbon balance, while the byproducts, such as O_3 and organic compounds were dramatically reduced, which was mainly due to increased amount of O formed from O_3 destruction [9]. Besides, plasma-driven catalysis has also been used in NO_x removal. It was reported that (plasma generated) ozone, hydroxyl radicals and atomic oxygen played important roles in the oxidation of NO to NO_2 [10] Many efforts have been made to purify VOC or NO_x gas by using plasma-driven catalysis [11]–[13]. However, those studies focused on only one specific polluted gas or some simulated gases, which were quite different from real vehicle exhaust gas from a launched car. In fact, complex components in the exhaust interacted with each other. For example, according to the molecular dynamics theory, NO removal efficiency depends heavily on the content of HC and O_2 [14]. Meanwhile, partially oxidized hydrocarbons and peroxy radicals (RO_2) will in return react with NO and strongly influence NO_2 formation rates [10]. Thus the application of the technique for examining the vehicle exhaust removal rate has a practical significance.

In previous studies, catalyst material can be introduced into the reactor in several ways, such as coating on the reactor wall or electrodes, as a packed-bed (granulates, coated fibers, pellets) or as a layer of catalyst material (powder, pellet, granulates, coated fiber) [8]. What researches worried about was the deactivation of catalyst [15], [16]. The catalyst contacted with the high electrode directly in all the studies above. What's more, most catalysts were prepared with liquid phase method, which contained too many complex chemical steps, even toxic or organic gas evaporating into the air [16], [17], [18], [19].

In this study, a novel and special structure plasma-driven catalysis device was proposed. The PDC reactor was a coaxial DBD reactor with three dielectric barrier layers and two copper electrodes. The middle dielectric barrier was designed to separate the catalyst from the high voltage electrode. The catalyst TiO_2, prepared by RF magnetron sputtering, was coated on the outer surface of the middle quartz tube. The electrodes surrounded the outer surfaces of corresponding dielectric barrier in a spiral way to prevent the damage to the catalyst for too much heat from microdischarges in PDC reactor. Then the PDC reactor was employed to treat the real vehicle exhaust. In order to explain the result, a simple and intuitive method–optical emission spectroscopy was conducted, which was different from chemical kinetics analysis in previous studies. This study also analyzed the practical problems for the application of PDC technique in real vehicle exhaust and proposed some solutions.

MATERIALS AND METHODS

The PDC Reactor

As shown in Fig. 1, the proposed PDC reactor was a coaxial DBD reactor with three quartz tubes as dielectric barrier layers and two copper electrodes. The inner electrode attached to the inner quartz tube with a spiral rotation, of which the helix width and the pitch were 6 mm and 2 mm, respectively, while the outer electrode attached to the outer quartz tube in the same way as the helix width of 9 mm and the pitch of 4 mm. The nano-titanium dioxide film prepared by radio frequency magnetron sputtering (JGP450 High vacuum magnetron sputtering, China) was coated on the outer surface of middle quartz tube. During the process of coating TiO_2 film, the Ti target was set at the bottom of the vacuum chamber to avoid a large amount of impurities groups into TiO_2 film, and the inner quartz rotated by a constant angular velocity of 0.15 r/s, guaranteeing a well-distributed film on the same circle of the reactor. The proportion of oxygen and argon was about 1:1 with the power of 150W at 1 Pa pressure. The XRD detection given inFig. 2 indicates that the titanium dioxide has an approximate composition of 85% anatase and 15% rutile forms of TiO_2. The discharge of the whole PDC reactor was shown in Fig. 3, with the excited voltage of 7.78 kV and the power of 2.75W. In order to know the surface features of the TiO_2, AFM detection was conducted on three points on the middle quartz tube with different distances of 20 mm, 50 mm, 90 mm from the same side, respectively. It was worth mentioning that the tube rotated at a certain speed of 0.15 r/s, which would help to form the uniform film in the same circle. Fig. 4 shows the AFM photograph of the uniform surface of the TiO_2. The length of the whole PDC reactor was 180 mm, and the outside diameter of the outer quartz tube was 18 mm, which constituted a compact reactor.

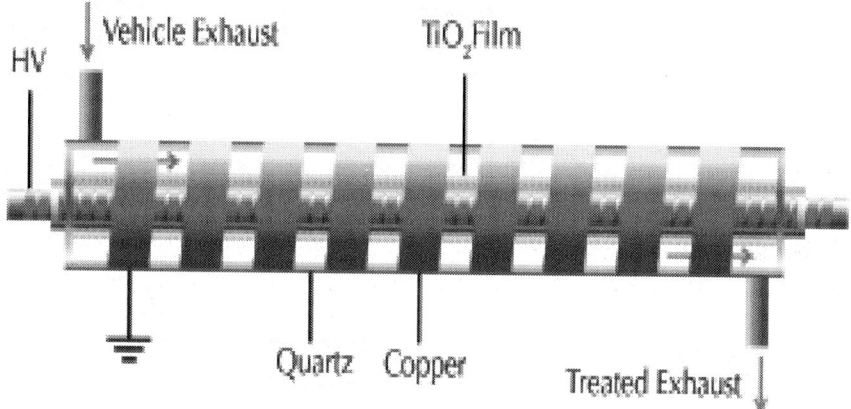

Figure 1: A schematic diagram of plasma-driven catalysis reactor. The PDC reactor was a coaxial DBD reactor with three quartz tubes as dielectric barrier layers and two copper electrodes. The outer and inner electrodes attached to the surfaces of the corresponding quartz tubes in a spiral rotation, respectively. The length of the whole reactor was 180 mm. The outside diameters of the quartz tubes were 18 mm, 10 mm and 6 mm, respectively.

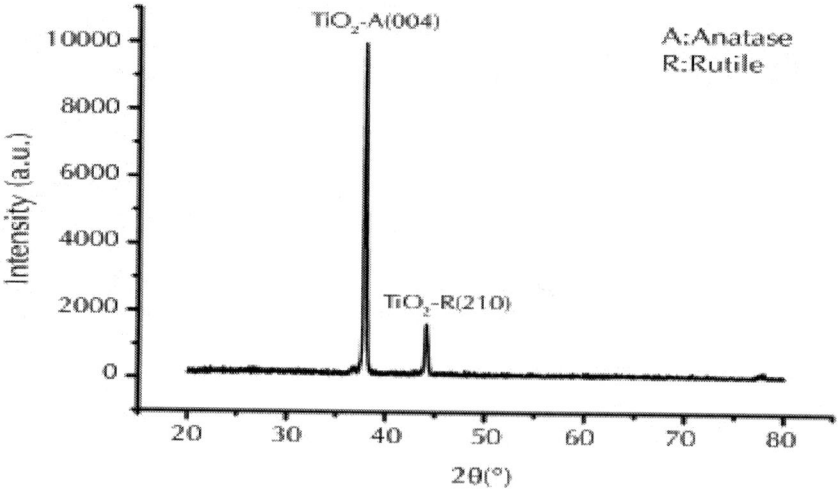

Figure 2: XRD pattern of prepared TiO_2 film. The titanium dioxide has an approximate composition of 85% anatase and 15% rutile forms of TiO_2.

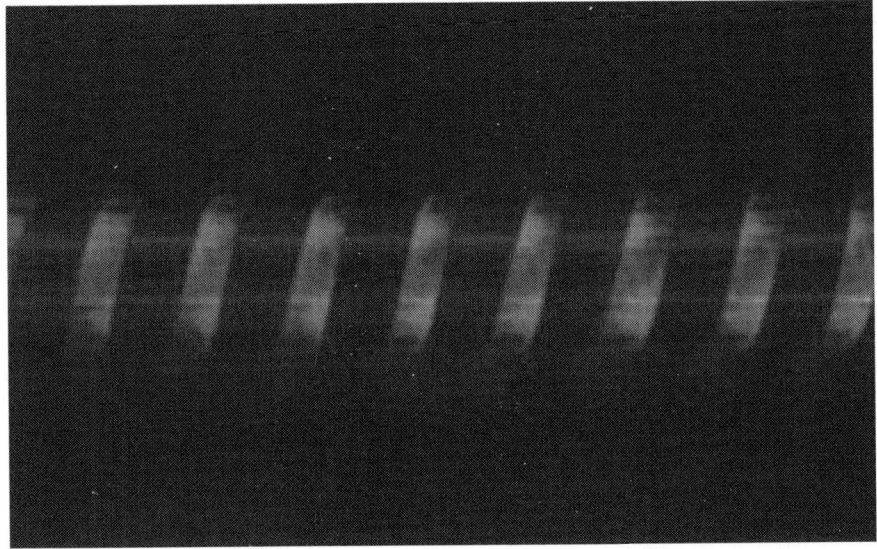

Figure 3: Discharge of plasma-driven catalysis reactor. The excited voltage and power were about 7.78 kV and 2.75W, respectively.

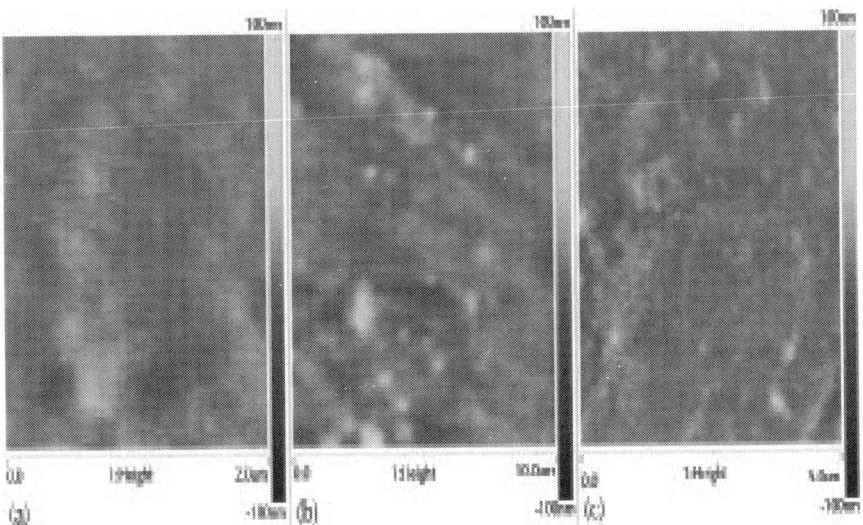

Figure 4: AFM photograph of prepared TiO_2 film. AFM detections on three points on the middle quartz tube with different distances of (a) 20 mm, (b) 50 mm and (c) 90 mm from the same side.

Clearance System

In this study, the experiment included exhaust gas source collection, treatment and components detection. The real exhaust gas was acquired from a jeep (Charokee-type Beijing Jeep 2500, made in China in 1999), in which the engine with inline four-cylinder was fed by #93 gasonline, in the condition of temperature/humidity 90°C/60% (monitored by MINGLE Hygrometer TH101B, China). The contents of components in the vehicle exhaust before and after treatment with the PDC reactor were detected by an exhaust gas analyzer (CV-5Q, Tianjin Shengwei Inc. Tianjin, China), which serves officially as a standard exhaust analysis in Beijing, China.

Clearance Process

At the beginning of each experiment, the jeep was started to let the vehicle exhaust gas steadily emit while the exhaust gas analyzer was turned on for detection. Then the exhaust was induced into the PDC reactor, which was excited for low temperature plasma later at the voltage of 8–10 kV. As shown by the detected spectrum in Fig. 5, many significant peaks can be observed in spectrum scope of working plasma inside the PDC reactor ranged from 250 nm to 520 nm, which satisfies the required wavelengths of TiO_2 catalysis.

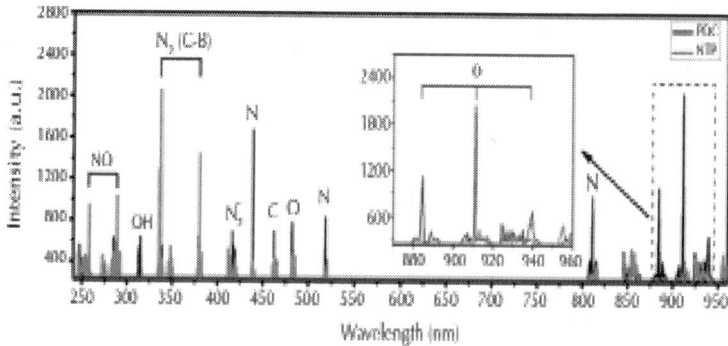

Figure 5: Spectrum of the discharge by PDC and NTP reactor. The appearance of O spectrum can be observed between 870 nm and 960 nm for wavelength in PDC reactor.

We then conducted experimental comparisons among three different groups, including a control group without any plasma treatment, a PDC group with TiO_2, and a NTP group which had an identical structure to the PDC reactor but without the catalyst TiO_2. Five tests were conducted for each group. In each test, after engine reaching steady state, the concentrations of different components of the vehicle exhaust were recorded sequentially at an interval of 20 s within 3 minutes for three groups. Then plasma was excited in both PDC and NTP reactors which the vehicle exhaust gas was lead into. The concentrations of different components were recorded at an interval of 20 s within another 9 minutes for these two groups.

Electrical Measurement

The PDC reactors were ignited by an AC high voltage power supply equipped with a transformer, which controlled the input power of the plasma generator (CTP-2000K). AC high frequency high voltage exported from the generator was applied to the PDC reactor, providing excitation power for the reactor. The excitation power was measured through the output voltage and current detection in the generator. Applied high voltage (V) was measured with a 1000:1 high voltage probe (TEKtronix, P6015A). V–Q Lissajous method was used to determine the discharge power in the PDC reactor. The charge Q was determined by measuring the voltage across the capacitor of 0.47 uF connected in series to the ground line of the PDC reactor. The voltage across this capacitor is proportional to the charge. The signals of applied voltage and charge were recorded with a digitizing oscilloscope (Tektronix, MSO2024) by averaging 62.5 k scans. The discharge power (P_{dis}) was evaluated from the area of V–Q parallelogram by multiplying the frequency. Specific input energy (SIE), which is defined as the energy input per unit gas-flow rate, can be obtained as follows:

$$SIE(J/L) = \frac{p_{dis}(watt)}{gas\,flow\,rate(L/\min)} \times 60 \tag{1}$$

In this study, both the minimum excited state and steady state of working PDC reactor were measured according to the method above.

Optical Emission Spectroscopy

Optical emission spectroscopy (OES) is one of the most widely used diagnostic methods for low-temperature plasmas [20]. Optical emission spectrometry involves applying electrical energy in the form of spark generated between an electrode and a metal sample, whereby the vaporized atoms are brought to a high energy state within a so-called "discharge plasma". These excited atoms and ions in the discharge plasma create a unique emission spectrum specific to each element. Thus, a single element generates numerous characteristic emission spectral lines. Therefore, the light generated by the discharge can be said to be a collection of the spectral lines generated by the elements in the sample. This light is split by a diffraction grating to extract the emission spectrum for the target elements. The intensity of each emission spectrum depends on the concentration of the element in the sample. Detectors (photomultiplier tubes) measure the presence or absence of the spectrum extracted for each element and the intensity of the spectrum to perform qualitative and quantitative analysis of the elements [21], [22]. The vehicle exhaust gas was induced into the PDC and NTP reactors, respectively. And the plasmas in the two reactors were excited with the same voltage and power. The probe was put at the same point of each reactor. Then the spectrums were obtained.

RESULTS

Vehicle Exhaust Clearance

Figure 6(a) and (b) show that both of the two reactors have a positive effect on the reduction of HC and NO. Moreover, the removal efficiency of PDC reactor was significantly higher than that of NTP reactor. The vehicle exhaust removal efficiency η was estimated as follow:

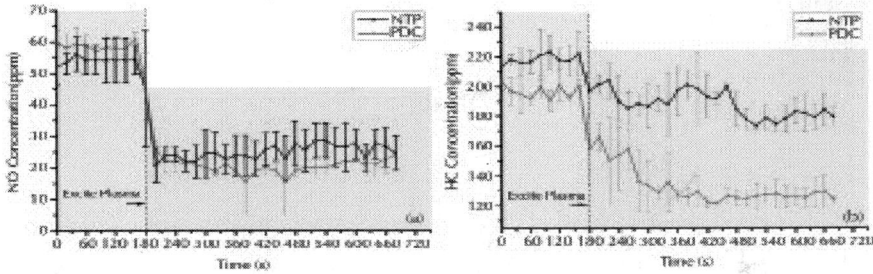

Figure6: Real-time removal results of (a) NO removal and (b) HC removal. The removal efficiency for HC in the PDC reactor was $\eta_{NO} = 64.5\%\pm1.8\%$, and the removal efficiency for NO in the PDC reactor was $\eta_{HC} = 32.1\%\pm1.3\%$

$$\eta(\%) = \frac{c_{in} - c_{out}}{c_{in}} \times 100\%$$

(2)

Where c_{in}, c_{out} are inlet and outlet concentration of the certain component, respectively.

According to Eq.1, the removal efficiency in the PDC reactor for HC and NO are as follows: $\eta_{NO} = 64.5\%\pm1.8\%$, and $\eta_{HC} = 32.1\%\pm1.3\%$.

Electrical Measurement

Experimental data indicated that our vehicle exhaust flow rate was about 590 sccm. The voltage and current of the PDC reactor were measured through capacitor sampling and resistance sampling, respectively. The V-Q Lissajous method was used to determine the discharge power. Both of the minimum excited state and steady state of PDC reactor were measured. All the electrical measurement results were shown in Fig. 7. At the minimum excited state, the effective voltage and current were 7.78 kV and 0.35 mA. Specific input energy was about 279.66 J/L. At the steady state, the effective voltage and current were 8.06 kV and 0.35 mA. Specific input energy was about 289.83 J/L.

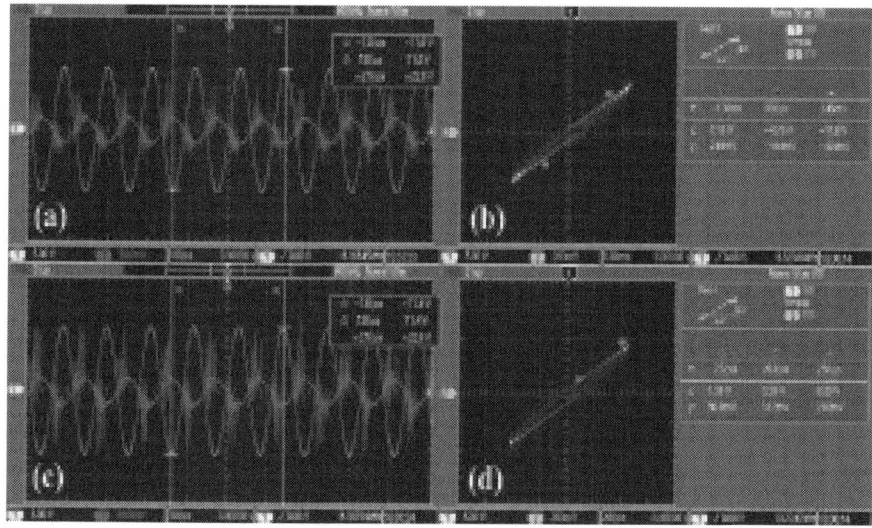

Figure 7: Electrical measurement of the PDC reactor. (a)The voltage and current of the PDC reactor at the minimum excited state. (b) Lissajous figure of the PDC reactor at the minimum excited state. (c) The voltage and current of the PDC reactor at the stable working state. (d) Lissajous figure of the PDC reactor at the stable working state.

Optical Emission Spectrum

The optical emission spectrum result was shown in Fig. 5. The OH, NO and N emission lines are visible. The appearance of O emission lines can be observed between 870 nm and 960 nm for wavelength in PDC reactor [23], [24], [25], [26]. Considering the facts that the removal efficiency of HC treated by PDC was much higher than that of NTP, while the removal efficiency of NO by PDC was only a little higher than that of NTP, we hypothesized that O might have made a significant contribution to the much higher HC and NO removal efficiencies.

DISCUSSION

First of all, in this study, the proposed PDC reactor was designed with three quartz tubes as dielectric layers. The outer and middle quartz tubes were both dielectric barrier layers. The middle one could separate the catalyst from the high voltage electrode, which could prevent high voltage electrode from being oxidized by oxygen during the reactions. On the other hand, the middle quartz tube could increase the area for depositing more TiO_2 film compared with a bare electrode. Furthermore, the addition of middle quartz tube could provide more chances to generate microdischarges [27], which would increase catalyst surface temperature [28], enhance the dispersion of active catalytic components [29], [30] and influence the stability with catalytic activity of the exposed catalyst material [31]. All above would promote catalytic VOC removal efficiency. However, hot spots can be formed in PDC reactors as a result of localizing heating by intense microdischarges, which might lead to the damage to the high voltage electrode and catalyst [32]. In order to avoid too many hot spots, each electrode surrounded the outer surfaces of corresponding dielectric barrier in a spiral way. The method of preparing TiO_2 film was RF magnetron sputtering without any toxic or organic gas evaporating into the air. Continuous operation tests indicated that stable performance without deterioration of catalytic activity could last for more than 25 h.

Secondly, the removal efficiency result was further explained through the optical emission spectrum approach, a simple and intuitive method different from chemical kinetics analysis in previous studies. The optical emission spectrum result showed that the appearance of O emission lines can be observed between 870 nm and 960 nm for wavelength in PDC reactor. It is our suggestion that the enhanced performance of hydrocarbon destruction was mainly due to a great amounts of atomic oxygen (O) formed, primarily from catalytically O_3 decomposition. Compared with NTP reactor, TiO_2 film acting as a semiconductor oxide catalyst provided a large number of free electron-hole pairs and consequently promoted the oxidation-reduction reaction.

The emission of O_3 from the NTP reactor was harmful to both human health and global environment. The addition of catalyst could significantly enhance the HC destruction with an increased O formation while the byproducts O_3 from the plasma were dramatically reduced [9]. Basically, O_3 formation in the NTP reactor proceeded via a two-step process [33]: formation of atomic oxygen and recombination of atomic oxygen with oxygen molecule (Eqs. (3)-(5)):

$$O_2 + electron \rightarrow O(^1D) + O(^3P)$$

(3

$$O(^1D) + M \rightarrow O(^3P) + M(M = O_2 or N_2)$$

(4)

$$O(^3P) + O_2 + M \rightarrow O_3 + M$$

(5)

Where O (^{1}D) and O (^{3}P) represent the excited and ground state oxygen atom, respectively.

It has been also reported that O_3 can be decomposed by catalysts into molecular oxygen via atomic oxygen and peroxides (Eqs. (6)- (8)) [34], where * denotes an active site on the catalyst surface:

$$O_3 + * \rightarrow O* + O_2$$

(6)

$$O* + O_3 \rightarrow O_2 + O_2 *$$

(7)

$$O_2 * \rightarrow O_2 + *$$

(8)

In general, atomic oxygen, which is highly active and involved in HC oxidation, is also imposed positive effect on NO destruction [9].

Thirdly, in the presence of catalyst TiO_2 film, when there is a faster rate of oxidation of hydrocarbons (Fig. 6b) there is not a significant increasing reduction of NO (Fig. 6a) correspondingly. Since the gas employed in this study was real vehicle exhaust gas containing many different kinds of compositions, there would be some reactions that did not occur in only one specific gas or some simulated gases. Actually, those complex components in the vehicle exhaust interacted with each other during the PDC process, which would remarkably influence the removal efficiency of NO and HC. It is worth mentioning that there exists a dynamic equilibrium between NO and HC, that HC decomposition will lead to the formation of NO [14]. Besides, it is known that both •OH and O_3 play an important role in NO removal. However, since O_3 is an •OH scavenger, partial O_3 and •OH will react with each other as follow [35]:

$$\bullet OH + O_3 \rightarrow O_2 + H_2O$$

(9)

Thus both contents of O_3 and •OH will decrease with an increased O_2 content. It has been reported in a previous literature that high O level will definitely lead to a conversion back to NO, and decrease the the removal efficiency of NO [36]. Although several studies have indicated that plasma-driven catalysis technique was quite effective in removing NO or hydrocarbon [37],[38], [39], [40], [41], according to our results and analysis above, the clearance rate of NO would be reduced if mixed with HC as well as other gas components.

The complex chemical reactions among gas compositions in the vehicle exhaust gas are briefly illustrated in Fig. 8. For example, higher NO removal efficiency is under the condition of lower content of HC or decreased O_2 content [14], meanwhile, NO can be removed by the formation of NO_2 through the reaction with partially oxidized hydrocarbons and peroxyl radicals (RO_2)[10]. Thus, the removal efficiency of HC could not be as high as that of NO in our experiment. Besides, many fundamental components from working plasma, including ozone, hydroxyl radicals and atomic oxygen also play an important role in the oxidation of NO to NO_2 [10]. When hydrocarbon was treated by the plasma discharge, partially oxidized hydrocarbons $(C_xH_yO_z)$ and peroxy radicals (RO_2) reacting with NO will be generated and strongly influenced NO_2 formation rate. Meanwhile, NO_2 reacted with the catalyst TiO_2 film, while partially oxidized hydrocarbons were consumed during selective catalytic reduction, producing CO_2, N_2, and H_2O, which are environment-friendly products [42]. Finally, •OH radicals can convert the formed NO_2 into HNO_3 [10] with the existence of H_2O [9]:

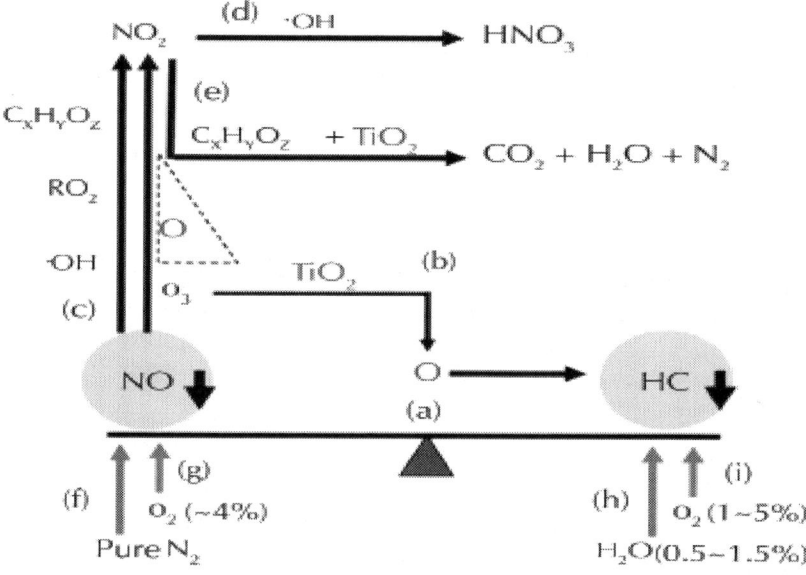

Figure 8: Chemical reactions gas compositions in the vehicle exhaust gas. (a)HC decomposition will lead to the formation of NO [14]. (b) The addition of catalysts could generate more single atomic oxygen from O_3 destruction, which contributed to the HC decomposition [9]. (c) single atomic oxygen, ozone, OH, oxidized hydrocarbons and peroxyl radicals played an important role in the oxidation of NO to NO_2 [10], and high O level will definitely lead to a conversion back to NO [36]. (d) •OH radicals can convert the formed NO_2 into HNO_3 [10] with the existence of H_2O [9]. (e) NO_2 reacted with the catalyst TiO_2 film, while partially oxidized hydrocarbons is consumed during selective catalytic reduction, producing CO_2, N_2, and H_2O. (f) Nitrogen in air can help keep the high removal efficiency of NO [45]. (g) NOx decomposition by plasma was known to be possible only if the oxygen content was less than about 4% [45], [46]. (h) The highest carbon balance and CO_2 selectivity for HC destruction were obtained with water vapor content between 0.5 and 1.5% [24]. (i) The optimal oxygen ranges between 1% and 5% for VOC with NTP [44].

$$e + H_2O \rightarrow e + \bullet OH + H \qquad (10)$$

$$O(^1D) + H_2O \rightarrow e + 2 \bullet OH. \qquad (11)$$

Based on the complex chemical reactions mentioned above, the removal efficiency of NO and HC should have been much higher if the PDC reactor was used to remove only NO or HC at the same amount of electricity consumption.

Besides, there were some other factors affecting the removal efficiency of NO and HC by using the proposed PDC reactor. On one hand, water vapor existing in the vehicle exhaust would reduce the vehicle exhaust removal efficiency. Although the drying modules had been assembled in our experiments, water vapor could not be removed completely and it would generate a considerable number of $\bullet OH$, leading to enhancement of NO conversion but decrease of HC removal efficiency with an increased incompletely oxidizing byproducts [10]. Meanwhile, water vapor can make catalyst deactivate through poisoning its active sites, annihilating high energetic electrons and depress the HC destruction through competing to be absorbed by the catalyst [43]. It has been reported that the highest carbon balance and CO_2 selectivity were obtained with water vapor content between 0.5 and 1.5% [17]. In further study, water vapor could be well controlled to further improve the exhaust removal efficiency.

On the other hand, similar to the presence of water vapor, the oxygen content in the vehicle exhaust that affects significantly the discharge performance plays a key role in the occurring chemical reactions. It has been mentioned that the optimal oxygen ranges between 1% and 5% for VOC with NTP [44], while NO_x decomposition by plasma was known to be possible only if the oxygen content was less than about 4% [45], [46]. In this study, the oxygen content in the exhaust ranged roughly from 3.5% to 4.2% all the time as shown in Fig. 9, which could benefit the PDC reactor for HC removal. However, the oxygen content above could have also benefited the PDC reactor for NO removal efficiency, but it was unreal for the exhaust gas containing VOC [14]. The effect of oxygen content on the NO and HC removal efficiency in this study was obviously reflected in the removal result as shown in Fig. 6.

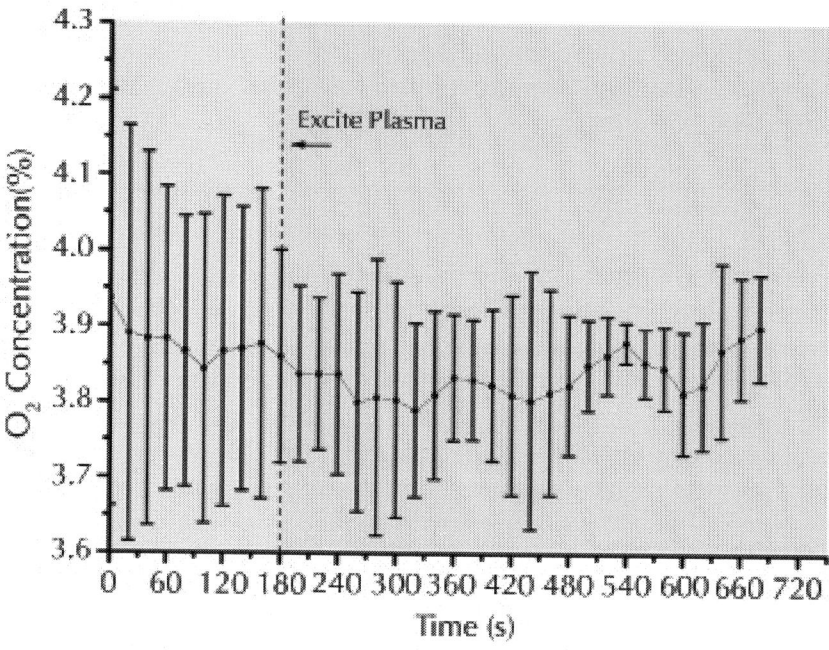

Figure 9: The O$_2$ content in the PDC reactor during the whole process. The oxygen content in the exhaust ranged roughly from 3.5% to 4.2% all the time.

In order to enhance the ability of the application of PDC reactor in vehicle exhaust gas, a reliable automatic control system will be expected. Based on the measured proportion of HC and NO in real-time, vehicle exhaust can be further removed effectively by the feedback control system that adaptively induces the air outside into PDC reactor. Particularly, the oxygen in air can be used to modulate the appropriate proportions among HC, NO and O$_2$ while the nitrogen in air can help keep the high removal efficiency of NO [45]. In this way, the real vehicle exhaust gas from cars will be well controlled with high efficiency in real-time.

ACKNOWLEDGMENTS

The support of Professor Zhang Guling and Ms. Wang Ke in Minzu University of China for preparing TiO$_2$ film is gratefully acknowledged.

AUTHOR CONTRIBUTIONS

Organized and helped with the design of the research: JZ JF. Conceived and designed the experiments: JZ. Performed the experiments: SY YL SS KZ. Analyzed the data: SY YL. Contributed reagents/materials/analysis tools: SY JF JZ KZ. Wrote the paper: SY JZ.

REFERENCES

1. Mista W, Kacprzyk R (2008) Decomposition of toluene using non-thermal plasma reactor at room temperature. Catal. Today 137: 345–349. doi: 10.1016/j.cattod.2008.02.009

2. Song CL, Bin F, Tao ZM, Li FC, Huang QF (2009) Simultaneous removals of NO_x, HC and PM from diesel exhaust emissions by dielectric barrier discharges. J. Hazard. Mater. 166: 523–349. doi: 10.1016/j.jhazmat.2008.11.068

3. Chen HL, Lee HM, Chen SH, Chang MB, Yu SJ, et al. (2009) Removal of Volatile Organic Compounds by Single-Stage and Two-Stage Plasma Catalysis Systems: A Review of the Performance Enhancement Mechanisms, Current Status, and Suitable Applications. Environ. Sci. Technol (43) 2216–2227. doi: 10.1021/es802679b

4. Fan X, Zhu T, Wang M, Li X (2009) Removal of low-concentration BTX in air using a combined plasma catalysis system. Chemosphere 75: 1301–1306. doi: 10.1016/j.chemosphere.2009.03.029

5. Francke KP, Miessner H, Rudolph R (2000) Plasmacatalytic processes for environmental problems. Catal. Today 59: 411–416. doi: 10.1016/s0920-5861(00)00306-0

6. Futamura S, Zhang A, Einaga H, Kabashima H (2002) Involvement of catalyst materials in nonthermal plasma chemical processing of hazardous air pollutants. Cataly. Today 72: 259–265. doi: 10.1016/s0920-5861(01)00503-x

7. Chen HL, Lee HM, Chen SH, Chao Y, Chang MB (2008) Review of plasma catalysis on hydrocarbon reforming for hydrogen production–Interaction, integration, and prospects. Appl. Phys. B: Environmental 85: 1–9. doi: 10.1016/j.apcatb.2008.06.021

8.   Durme JV, Dewulf J, Leys C, Langenhove HV (2008) Combining non-thermal plasma with heterogeneous catalysis in waste gas treatment: A review. Appl. Phys. B: Environmental 78: 324–333. doi: 10.1016/j.apcatb.2007.09.035

9.   Huang HB, Ye DQ, Dennis YCL (2011) Abatement of Toluene in the Plasma-Driven Catalysis: Mechanism and Reaction Kinetics. IEEE Trans. Plasma Sci. 39: 877–882. doi: 10.1109/tps.2010.2103403

10.   Maciuca A, Dupeyrat CB, Tatibouët JM (2008) Synergetic effect by coupling photocatalysis with plasma for low VOCs concentration removal from air. Appl. Catal. B: Environ. 125: 432–438. doi: 10.1016/j.apcatb.2012.06.012

11.   Rappe KG, Hoard W, Aardahl CL, Park PW, Peden CHF, et al. (2004) Combination of low and high temperature catalytic materials to obtain broad temperature coverage for plasma-facilitated NO_x reduction. Catal. Today 89: 143–150. doi: 10.1016/j.cattod.2003.11.020

12.   Oh SM, Kim HH, Einaga H, Ogata A, Futamura S, et al. (2006) Zeolite-combined plasma reactor for decomposition of toluene. Thin Solid Films 506: 418–422. doi: 10.1016/j.tsf.2005.08.105

13.   Holzer F, Roland U, Kopinke FD (2002) Combination of non-thermal plasma and heterogeneous catalysis for oxidation of volatile organic compounds: Part 1. Accessibility of the intra-particle volume. Appl. Catal. B: Environ. 32: 163–181. doi: 10.1016/s0926-3373(02)00040-1

14.   Kima HH, Ogat A (2011) Nonthermal plasma activates catalyst: from current understanding and future prospects. Phys. J. Appl. Phys. 55, 13806.

15.   Harling AM, Kim HH, Futamura S, Whitehead J C (2007) Temperature Dependence of Plasma-Catalysis Using a Nonthermal, Atmospheric Pressure Packed Bed; the Destruction of Benzene and Toluene. J. Phys. Chem. C 2007, 111, 5090–5095.

16.   Kim HH, Oh SM, Ogata A, Futamura S (2004) Decomposition of benzene using Ag/TiO2 packed plasma-driven catalyst reactor: influence of electrode configuration and Ag-loading amount. Catalysis Letters Vol. 96, Nos. 3–4, July.

17. Huang HB, Ye DQ, Dennis YCL (2011) Plasma-Driven Catalysis Process for Toluene Abatement: Effect of Water Vapor. IEEE Trans. Plasma Sci. 39: 576–580. doi: 10.1109/tps.2010.2086498

18. Ding HX, Zhu AM, Lu FG, Xu Y, Zhang J, et al. (2006) Low-temperature plasma-catalytic oxidation of formaldehyde in atmospheric pressure gas streams. J. Phys. D: Appl. Phys. 39: 3603–3608. doi: 10.1088/0022-3727/39/16/012

19. Harling AM, Demidyuk V, Fischer SJ, Whitehead JC (2008) Plasma-catalysis destruction of aromatics for environmental clean-up: Effect of temperature and configuration. Applied Catalysis B: Environmental 82: 180–189. doi: 10.1016/j.apcatb.2008.01.017

20. Zhu X, Pu Y (2010) Optical emission spectroscopy in low-temperature plasmas containing argon and nitrogen: determination of the electron temperature and density by the line-ratio method. J. Phys. D: Appl. Phys. 43: 403001. doi: 10.1088/0022-3727/43/40/403001

21. Fantz U (2006) Basics of plasma spectroscopy. Plasma Sources Sci. Technol. 15: S137–S147. doi: 10.1088/0963-0252/15/4/s01

22. Staack D, Farouk B, Gutsol AF, Fridman AA (2006) Spectroscopic studies and rotational and vibrational temperature measurements of atmospheric pressure normal glow plasma discharges in air. Plasma Sources Sci. Technol. 15 (2006) 818–827.

23. Walsh JL, Kong MG (2008) Contrasting characteristics of linear-field and cross-field atmospheric plasma jets. Appl. Phys. Lett. 93, 111501.

24. Lee YH, Yi CH, Chung MJ, Yeom GY (2001) Characteristics of He/O_2 atmospheric pressure glow discharge and its dry etching properties of organic materials. Surf. Coat. Technol. 146–147: 474–479. doi: 10.1016/s0257-8972(01)01459-1

25. Lofthus A, Krupenie PH (1977) The spectrum of molecular nitrogen. J. Phys. Chem. Ref. Data 6, 113.

26. Xu L, Nonaka H, Zhou HY, Ogino A, Nagata T, et al. (2007) Characteristics of surface-wave plasma with air-simulated N_2–O_2 gas mixture for low-temperature sterilization. J. Phys. D: Appl. Phys. 40: 803–808. doi: 10.1088/0022-3727/40/3/017

27. Kogelschatz U (2003) Dielectric-barrier Discharges: Their History, Discharge Physics, and Industrial Applications. Plasma Chem. Plasma Process. Vol. 23, No. 1, March.

28. Lu B, Zhang X, Yu X, Feng T, Yao S (2006) Catalytic oxidation of benzene using DBD corona discharges, Journal of Hazardous Materials. 137: 633–637. doi: 10.1016/j.jhazmat.2006.02.012

29. Guo YF, Ye DQ, Chen KF, He JC, Chen WL (2006) Toluene decomposition using a wire-plate dielectric barrier discharge reactor with manganese oxide catalyst in situ. Journal of Molecular Catalysis A: Chemical 245: 93–100. doi: 10.1016/j. molcata.2005.09.013

30. Zhang YP, Ma PS, Zhu XL, Liu CJ, Shen YT (2004) A novel plasma-treated Pt/NaZSM-5 catalyst for NO reduction by methane, Catalysis Communications. 5: 35–39. doi: 10.1016/j. catcom.2003.11.006

31. Guo YF, Ye DQ, Chen KF, He JC (2007) Toluene removal by a DBD-type plasma combined with metal oxides catalysts supported by nickel foam, Catalysis.

32. Kim HH, Ogata A, Futamura S (2006) Effect of different catalysts on the decomposition of VOCS using flow-type plasma-driven catalysis, IEEE Transactions on Plasma Science. 34: 984–995. doi: 10.1109/tps.2006.875728

33. H. H Kim, A Ogata, S Futamura (2006) Effect of different catalysts on the decomposition of VOCs using flow-type plasma-driven catalysis, IEEE Trans. Plasma Sci. 34: 984–995. doi: 10.1109/ tps.2006.875728

34. Futamura S, Einaga H, Kabashima H, Hwan LY (2004) Synergistic effect of silent discharge plasma and catalysts on benzene decomposition. Catal. Today 89: 89–95. doi: 10.1016/j. cattod.2003.11.014

35. Kim HH, Oh SM, Ogata A, Futamura S (2005) Decomposition of gas-phase benzene using plasma-driven catalyst (PDC) reactor packed with Ag/TiO2 catalyst. Appl. Catal. B Environ (56) 213–220. doi: 10.1016/j.apcatb.2004.09.008

36. Malik MA, Kolb JF, Sun Y, Schoenbach KH (2011) Comparative study of NO removal in surface-plasma and volume-plasma reactors based on pulsed corona discharges. J. Hazard. Mater. 197: 220–228. doi: 10.1016/j.jhazmat.2011.09.079

37. X Tu, J.C Whitehead (2012) Plasma-catalytic dry reforming of methane in an atmospheric dielectric barrier discharge:

Understanding the synergistic effect at low temperature. Appl. Catal. B. 125: 439–448. doi: 10.1016/j.apcatb.2012.06.006

38. Fan HY, Shi C, Li XS, Zhao DZ, Xu Y, et al. (2009) High-efficiency plasma catalytic removal of dilute benzene from air. J. Phys. D: Appl. Phys. 42: 225105. doi: 10.1088/0022-3727/42/22/225105

39. Ding HX, Zhu AM, Lu FG, Xu Y, Zhang J, et al. (2006) Low-temperature plasma-catalytic oxidation of formaldehyde in atmospheric pressure gas streams. J. Phys. D: Appl. Phys. 39: 3603–3608. doi: 10.1088/0022-3727/39/16/012

40. Mok YS, Koh DJ, Kim KT, Nam IS (2003) Nonthermal Plasma-Enhanced Catalytic Removal of Nitrogen Oxides over V_2O_5/TiO_2 and Cr_2O_3/TiO_2. Ind. Eng. Chem. Res., 42 (13): 2960–2967. doi: 10.1021/ie0208873

41. Chen Z, Mathur VK (2003) Nonthermal Plasma Electrocatalytic Reduction of Nitrogen Oxide. Ind. Eng. Chem. Res. 42 (26): 6682–6687. doi: 10.1021/ie030096g

42. Lin H, Huang Z, Shangguan W, Peng X (2007) Temperature-programmed oxidation of diesel particulate matter in a hybrid catalysis–plasma reactor. Proc. Combust. Inst 31: 3335–3342. doi: 10.1016/j.proci.2006.07.075

43. Zhang PY, Liang FY, Yu G, Chen Q, Zhu WP (2003) A comparative study on decomposition of gaseous toluene by O_3/UV, TiO_2/UV and $O_3/TiO_2/UV$. J. Photochem. Photobiol. A: Chemistry 156: 189–194. doi: 10.1016/s1010-6030(02)00432-x

44. Vandenbroucke AM, Morent R, Geyter ND, Leys C (2011) Decomposition of Trichloroethylene with Plasma-catalysis: A review. J. Hazard. Mater. 195: 30–54.

45. Masuda S, Hosokawa S, Tu X, Sakakibara K, Kitoh S (1993) et.al (1993) Destruction of gaseous pollutants by surface-induced plasma chemical process (SPCS). IEEE Trans. Ind. Applicat. 29: 781–786. doi: 10.1109/28.231994

46. Yan K, Kanazawa S, Ohkubo T, Nomoto Y (1999) Oxidation and Reduction Processes During NO_x Removal with Corona-Induced Nonthermal Plasma. Plasma Chem. Plasma Proc. 19: 421–443. doi: 10.1023/a:1021824504271.

Effect of Inlet and Outlet Flow Conditions on Natural Gas Parameters in Supersonic Separation Process

Yan Yang[1], Chuang Wen[1], Shuli Wang[1], and Yuqing Feng[2]

[1]Jiangsu Key Laboratory of Oil-Gas Storage and Transportation Technology, Changzhou University, Changzhou, Jiangsu Province, China

[2]Computational Informatics, Commonwealth Scientific and Industrial Research Organization, Melbourne, the State of Victoria, Australia

ABSTRACT

A supersonic separator has been introduced to remove water vapour from natural gas. The mechanisms of the upstream and downstream influences are not well understood for various flow conditions from the wellhead and the back pipelines. We used a computational model to investigate the effect of the inlet and outlet flow conditions on the supersonic separation process. We found that the shock wave was sensitive to the inlet or back pressure compared to the inlet temperature. The shock position shifted forward with a higher inlet or back pressure. It indicated that an increasing inlet pressure declined the pressure recovery capacity. Furthermore, the shock wave moved out of the diffuser when the ratio of the back pressure to the inlet one was greater than 0.75, in which the state of the low pressure and temperature was destroyed, resulting in the re-evaporation of the condensed liquids. Natural gas would be the subsonic flows in the whole supersonic separator, if the mass flow rate was less than the design value, and it could not reach the low pressure and temperature for the condensation and separation of the water vapor. These results suggested a guidance mechanism for natural gas supersonic separation in various flow conditions.

INTRODUCTION

As the global economy rises, the demand for energy supply is increasing continuously in the last two decades. Natural gas plays a significant strategic role in the energy supply [1]. Natural gas is gaseous mixture, primarily composed of methane, ethane, propane and butane, with some heavier alkanes, carbon dioxide, hydrogen sulfide, nitrogen and a small amount of water vapor [2]. The presence of water vapor in natural gas increases the risk of the formation of gas hydrates with line plugging due to hydrate deposition on the pipe walls, results in corrosion combined with acid gases including carbon dioxide and hydrogen sulfide, and reduces the delivery capacity of the pipelines because of the collection of free water [3]. Consequently, the water vapor must be removed from natural gas early on.

At present, many conventional techniques are employed for the natural gas separation, such as absorption, adsorption, refrigeration, membranes and so on. A supersonic separator, as a novel technique, has been introduced to natural gas processing from the beginning of this century[4]–[6]. In essence, the supersonic separation technique causes refrigeration like the Joule-Thompson effect and Turbine expansion, both of which induce a low temperature for the condensation of water vapor. The supersonic separator mainly consists of a Laval nozzle, a swirl device and a diffuser.

Malyshkina [7], [8] obtained the distribution of gas dynamic parameters through a supersonic separator with a computational method, and a procedure was developed to predict the separation capability of water vapor and higher hydrocarbons from natural gas by using a supersonic separator determined by the initial parameters. Karimi and Abdi [9] studied the flow fields of natural gas in a Laval nozzle of 0.12 m long. But the working fluid was assumed to be a supercritical flow. The geometric construction and flow conditions are quite different from the actual flow states of natural gas in a supersonic separator for dehydration. Jiang et al. [10]employed the corrected Internally Consistent Classical Theory and Gyarmathy theory to modelling the nucleation and droplet growth of natural gas in the supersonic separation process. A supersonic separator was compared to a Joule-Thomson valve with TEG and the results demonstrated the high economic performance and natural gas liquids recovery of a supersonic separator [11]. The generalized radial basis function artificial neural networks were used to optimize the geometry of a supersonic separator [12]. Rajaee Shooshtari and Shahsavand developed a new theoretical approach based on mass transfer rates to calculate the liquid droplet growth in supersonic conditions for binary mixtures [13]. In our preliminary studies, a central body was incorporated in a supersonic separator with a swirling device composed of vanes and an ellipsoid [14]. The effects of swirls on natural gas flow in supersonic separators were computationally simulated with the Reynolds stress model [15]. The particle separation characteristic in a supersonic separator was calculated using the discrete particle method [16].

The mechanisms of the upstream and downstream influences are not well understood for various flow conditions from the wellhead and the back pipelines. The purpose of this study is to investigate the effects of the operating parameters on natural gas supersonic separation process, including the back pressure, inlet mass flow rates, inlet pressures and inlet temperatures. The Redlich-Kwong real gas model is employed to calculate the gas thermal properties in high pressure and low temperatures in our simulation.

GOVERNING EQUATIONS

Natural gas can be accelerated to supersonic velocities with a Laval nozzle in a supersonic separator and, accordingly, low pressure and temperature conditions are achieved for water vapor condensation. The fluid structure of natural gas flows can be described by the conservation equations of mass, momentum and energy. To close the partial differential equations, the Shear Stress Transport (SST) [17] turbulence model was used in our simulation to solve the supersonic gas flows.

The mass equation of gas phase (continuity equation) is described as:

$$\frac{\partial}{\partial x_i}(\rho u_i) = 0$$

(1)

Where and u are the gas density and velocity, respectively.

The conservation of momentum for gas phase can be written as follows:

$$\frac{\partial}{\partial x_j}(\rho u_i u_j + p\delta_{ij} - \tau_{ji}) = 0$$

(2)

Where p is the gas pressure; τ_{ij} is the viscous stress; δ_{ij} is the Kronecker delta.

The energy equation for gas phase is expressed as Eq.

$$\frac{\partial}{\partial x_j}(\rho u_j E + u_j p + q_j - u_i \tau_{ij}) = 0$$

(3)

Where E is the total energy; q_j is the heat flux; t is the time.

The turbulent kinetic energy and the specific dissipation rate equations in SST model are as follows [17], [18]:

$$\frac{\partial}{\partial x_i}(\rho k u_i) = \frac{\partial}{\partial x_j}\left(\Gamma_k \frac{\partial k}{\partial x_j}\right) + \bar{G}_k - Y_k + S_k$$

(4)

$$\frac{\partial}{\partial x_j}(\rho \omega u_j) = \frac{\partial}{\partial x_j}\left(\Gamma_\omega \frac{\partial \omega}{\partial x_j}\right) + G_\omega - Y_\omega + D_\omega + S_\omega$$

(5)

Where k is the turbulent kinetic energy, ω is the specific dissipation rate. Γ_k and Γ_ω. represent the effective diffusivity of k and ω, respectively. $\overline{G_k}$ represents the generation of turbulence kinetic energy due to mean velocity gradients. G_ω represents the generation of the specific dissipation rate, ω. Y_k and Y_ω represent the dissipation of k and ω due to turbulence. D_ω represents the cross-diffusion term. S_k and S are user-defined source terms.

An equation of state must be developed to calculate the physical property of fluids in supersonic flows. In this simulation, the Redlich-Kwong real gas equation of state model [19] was employed to predict gas dynamic parameters, described in Eq. (6).

$$p = \frac{RT}{V_m - b} - \frac{a}{\sqrt{T}V_m(V_m + b)}$$

(6)

Where p is the gas pressure, R is the gas constant, T is temperature, V_m is the molar volume (V/n), a is a constant that corrects for attractive potential of molecules, and b is a constant that corrects for volume.

The constants a and b are different depending on which gas is being analyzed. They can be calculated from the critical point data of the gas:

$$a = \frac{0.4275 R^2 T_c^{2.5}}{p_c}$$

(7)

$$b = \frac{0.08664 R T_c}{p_c}$$

(8)

Where T_c and p_c are the temperature and pressure at the critical point, respectively.

For the multi-component mixtures, such as natural gas, mixing laws are utilized to calculate the parameters a and b. The Van Der Waals mixing rules [20], [21] were applied to obtain the parameters for the mixtures from those pure components. The mathematical expressions of this mixing rule can be written,

$$a = \sum_{i=1}^{n} \sum_{j=1}^{n} x_i x_j \sqrt{a_i a_j} (1 - k_{ij})$$

(9)

$$b = \sum_{i=1}^{n} x_i b_i$$

(10)

Where x is molar fraction; n is the total number of the gas components; k_{ij} is the binary interaction parameter between components i and j.

MATHEMATICAL MODELLING

Computational Domain and Boundary Conditions

A Laval nozzle is a key part of a supersonic separator to generate supersonic flows for the condensation and separation of natural gas. Thus, the nozzle needs to be designed specifically, as shown in Figure 1. The cubic polynomial equation was employed to calculate the converging contour of the nozzle, as shown in Eq. (11), while the Foelsch's analytical calculation method was used to design the diverging part of the nozzle [22]. This design of the converging part will accelerate the gas flow uniformly to achieve the sound speed in the throat area. The critical cross-section area is 0.0002378 m². The nozzle entrance and exit areas are 0.007854 m² and 0.0004460 m², respectively. In addition, a straight tube with the length of 100 mm was connected to the nozzle upstream and diffuser downstream, respectively.

$$\begin{cases} \dfrac{D-D_{cr}}{D_1-D_{cr}} = 1 - \dfrac{1}{X_m{}^2}\left(\dfrac{x}{L}\right)^3 & \left(\dfrac{x}{L} \le X_m\right) \\[3mm] \dfrac{D-D_{cr}}{D_1-D_{cr}} = \dfrac{1}{(1-X_m)^2}\left(1-\dfrac{x}{L}\right)^3 & \left(\dfrac{x}{L} > X_m\right) \end{cases}$$

(11)

Where D_1, D_{cr} and L are the inlet diameter, the throat diameter and the convergent length, respectively. $X_m = 0.45$. x is the distance between arbitrary cross section and the inlet, and D is the convergent diameter at arbitrary cross section of x.

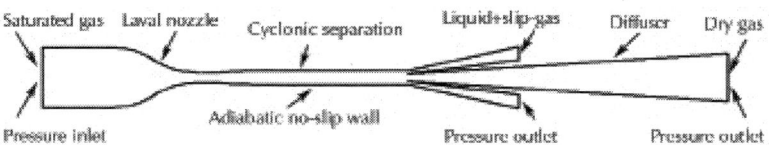

Figure 1: Schematic diagram of a supersonic separator.

A structured grid was generated for the supersonic separator while a finer grid scheme in the boundary layer was employed in Laval nozzle and supersonic channel. The grid independence was tested before we carried out the simulation. Boundary conditions played a significant role in a numerical simulation. In our case related to a supersonic separator, the pressure boundary conditions were assigned for the inlet and outlet of the supersonic separator, respectively, according to the flow characteristics of the supersonic compressible fluid,. No-slip and adiabatic boundary conditions were specified for the walls. The turbulent kinetic energy and turbulent dissipation rate were employed as the turbulence parameters.

Computational Methods

The finite volume methods were used to discretize the partial differential equations of the supersonic gas flows. The pressure based implicit solver was employed to solve the governing equations. The SIMPLE algorithm [23] was applied to couple the velocity field and pressure. The standard pressure scheme was adopted to interpolate the pressure values on the surface of the control volume. The second-order upwind scheme was used for other variables, such as density, momentum, turbulence kinetic energy, turbulence dissipation rate.

Validation

For the validation of our computational methods in supersonic flows, it was validated with Arina's results before we applied it to our designed supersonic separator [24], [25]. Figure 2depicts the pressure profiles in a Laval nozzle with the numerical results and Arina's work. It could be seen that the same flow behavior was obtained and the shock wave position was accurately captured by our simulation method. Therefore, the numerical results agree with Arina's results well. It was demonstrated that our developed model could be used in the prediction of the supersonic flow for natural gas dehydration.

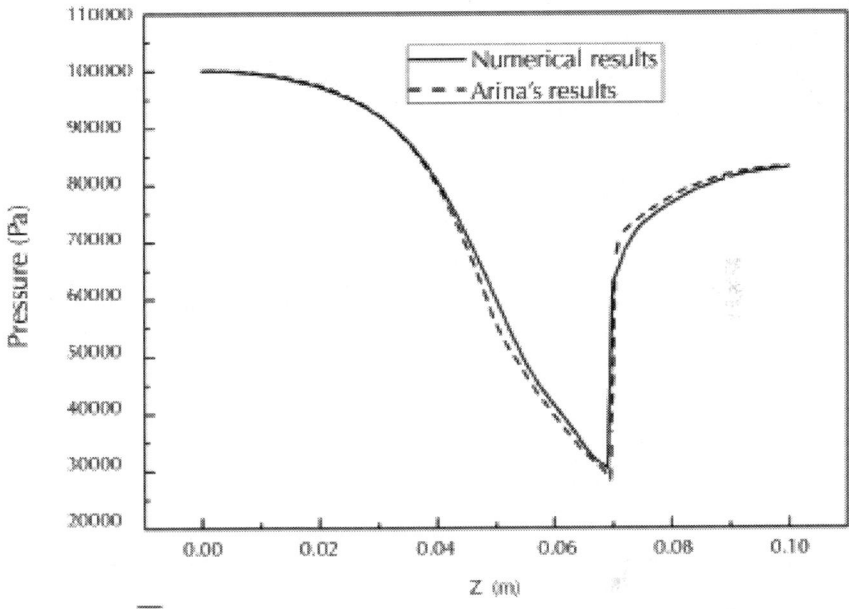

Figure 2: Pressure profile for nozzle flow.

RESULTS AND DISCUSSION

Effect of Back Pressure

The flow characteristics of natural gas were numerically simulated in the supersonic separation process. The multi-components gas mixture in Baimiao gas well of Zhongyuan Oil Field was selected for the calculation. The composition of natural gas in mole fraction is shown in Table 1.

Table 1: Mole composition of natural gas

Natural gas composition	Mole fraction (%)

CH4	91.36
C2H6	3.63
C31-18	1.44
C4H10	0.26
n-C4H10	0.46
i-C5H12	0.17
n-C5H12	0.16
H2O	0.03
CO2	0.45
N2	2.04

The incoming flow parameters are fixed when we study the effect of the back pressures on the supersonic separation process. The detailed initial conditions for the back pressure simulation are shown in Table 2. Figure 3 presents the static pressure and static temperature profiles along the flow direction in the conditions of different back pressures. The shock wave position moves into the nozzle from the diffuser with the rise of the back pressure. The shock wave will stay in the diffuser while the back pressure is about less than 75 bar with the inlet pressure of about 100 bar. If the back pressure increases to 80 bar, the shock wave will move into the supersonic channel across the diffuser entrance. The pressure and temperature profiles exhibit several fluctuations close to the shock wave and away from it. This is induced by the interaction between the boundary layer separation and the shock boundary layer.

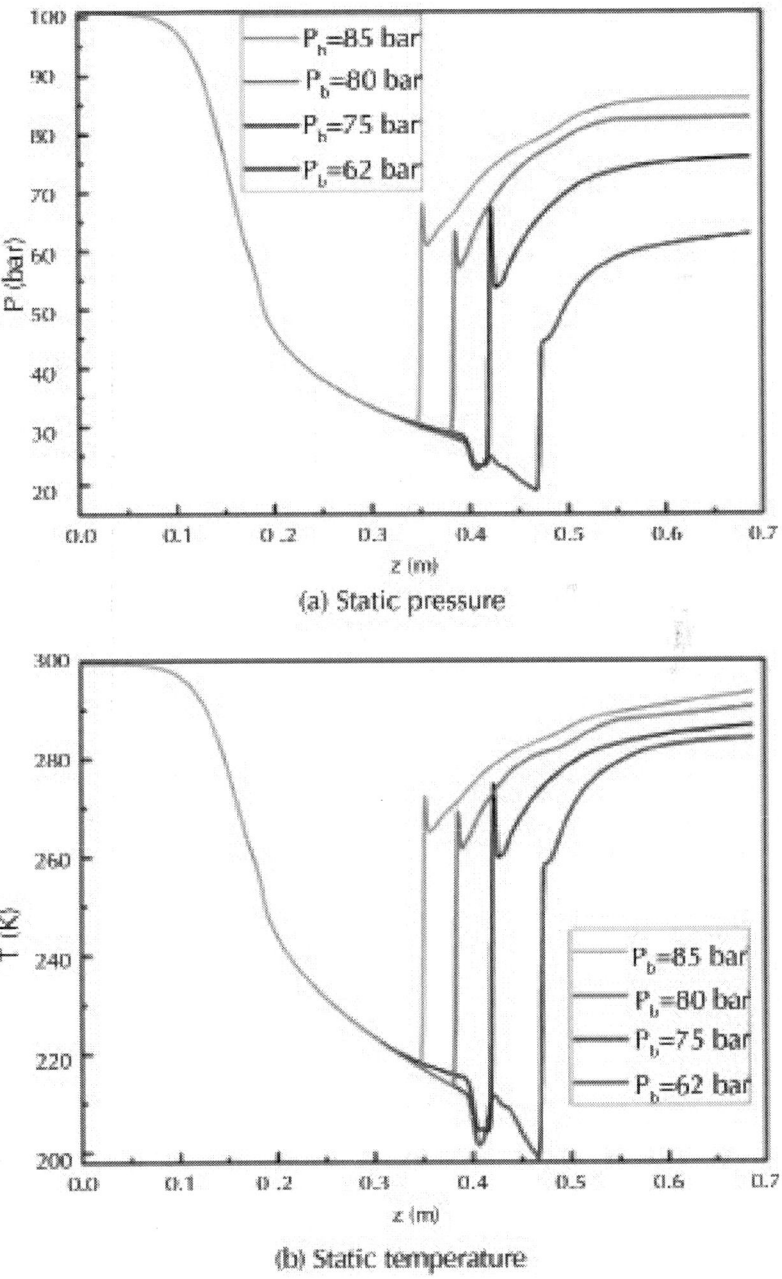

(a) Static pressure

(b) Static temperature

Figure 3: Effect of back pressure on natural gas dynamic parameters.

Table 2: Initial conditions for back pressure simulation

Cases	Inlet pressure (bar)	Inlet temperature (K)	Back pressure (bar)
1	100	300	85
2	100	300	80
3	100	300	75
4	100	300	62

Figure 4 depicts the contours of gas Mach numbers in the supersonic separators with various back pressures. It clearly shows the obvious differences of the shock wave position with the increasing back pressure. In this simulation case, the shock wave even goes into the nozzle diverging part when the back pressure reaches 85 bar. In this condition, the shock wave will destroy the state of the low pressure and temperature, resulting in the re-evaporation of the condensed liquids to decline the separation efficiency of the supersonic separators.

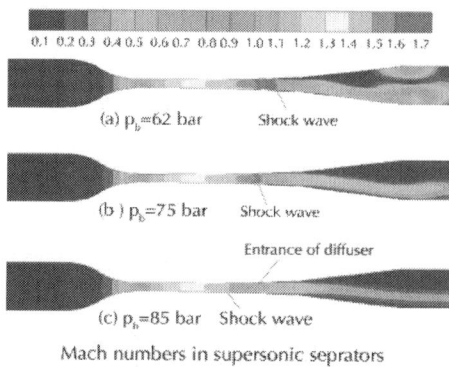

Mach numbers in supersonic seprators

Figure 4: Mach numbers in supersonic separators with various back pressures.

Effect of Inlet Mass Flow Rate

A Laval nozzle is a key part in a supersonic separator, and the critical area at the nozzle throat determines the gas mass flow rate through this device. The detailed initial conditions for inlet mass flow rate simulation are shown in Table 3. Figure 5 describes the gas dynamic parameters with various inlet mass flow rate, namely, including the gas Mach number, the static pressure and static temperature. If the mass flow rate is less than the design value, the gas velocity at the nozzle throat is less than the critical value, although the converging part speeds up the gas flows. Because of the Mach number at the throat is less than unity, the gas velocity declines in the diverging part of the Laval nozzle. In this situation, the maximum velocity is obtained at the nozzle throat. That is, natural gas is the subsonic flows in the whole supersonic separator, which cannot reach the low pressure and temperature for the condensation of the water vapor. The gas Mach number rises with the increase of the inlet gas mass flow rate, resulting in the decline of the static pressure and temperature. When the inlet gas flow rate reaches the design value, the choked flow conditions will be achieved. In our simulation cases, the critical flow condition is obtained when the inlet gas mass flow rate is about 4.155 kg/s. In this condition, the natural gas flow continues to expand in the diverging part of the Laval nozzle, and the maximum Mach number is around 1.33.

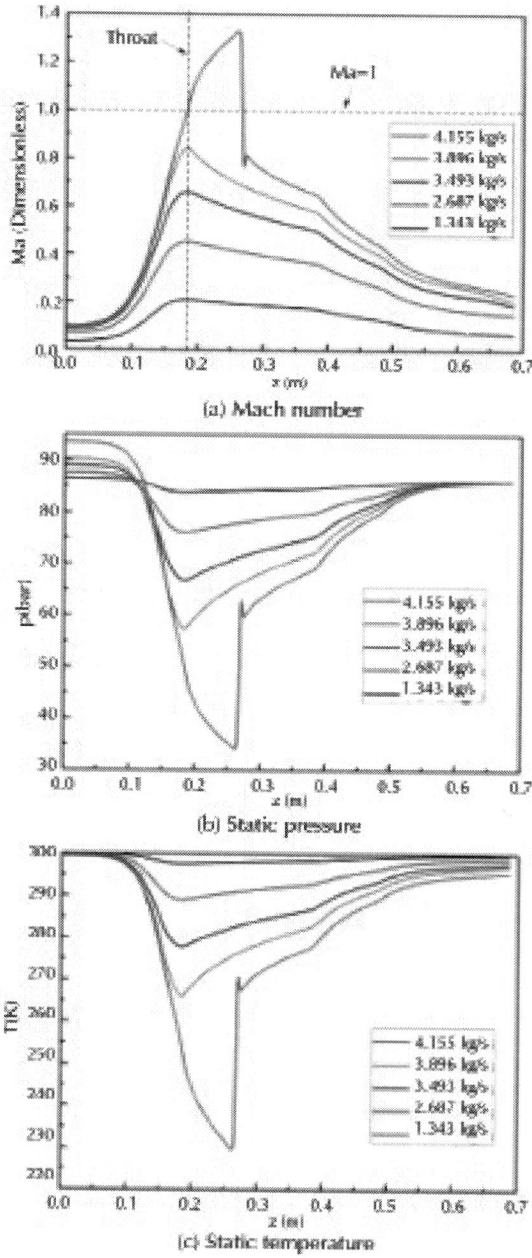

Figure 5: Effect of inlet mass flow rate on natural gas dynamic parameters.

Table 3: Initial conditions for inlet mass flow rate simulation

Cases	Inlet mass flow rate (kg/s)	Inlet temperature (K)	Back pressure (bar)
1	1.343	300	85
2	2.687	300	85
3	3.493	300	85
4	3.896	300	85
5	4.000	300	85

Figure 6 depicts the phase envelope curve and the pressure–temperature (P-T) profiles with various inlet mass flow rates. We can see that P-T profile doesn't reach the phase envelope curve because of the high pressure and temperature in the supersonic separator, when the inlet gas mass flow rate is smaller than the design value. Therefore, the water vapor can hardly be removed from natural gas when the inlet mass flow rate is less than the designed rate.

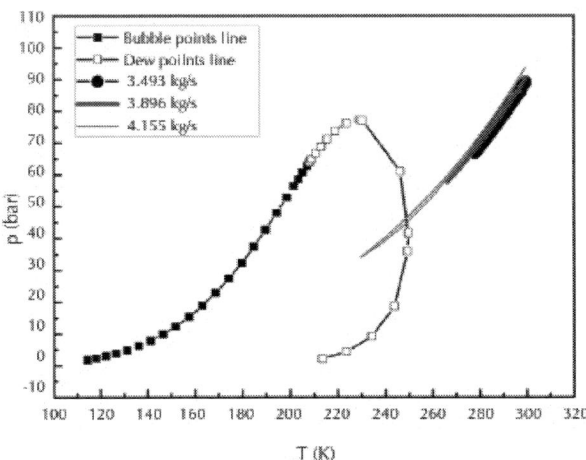

Figure 6: Phase envelope and pressure–temperature relationships with various inlet mass flow rates.

Effect of Inlet Pressure

The inlet temperature is fixed and the back pressure is set to be the 85% of the inlet one, when we studied the effect of the inlet pressures on the gas dynamic parameters. The detailed initial conditions for inlet pressure simulation are shown in Table 4. The gas mass flow rate in a supersonic separator increases with the rises of the inlet pressure. It indicates that the processing capacity of a supersonic separator can be improved by increasing the inlet pressure in natural gas processing. Figure 7 presents the gas static pressure and temperature profiles along the designed supersonic separator. The shock wave position shifts forward to the nozzle with a higher inlet pressure. For example, the shock position stayed at z = 0.370 m at an inlet pressure of about 50 bar. However, the shock location goes to the upstream of the nozzle divergent part, at z = 0.263 m, when the inlet increases to 300 bar. That is, the shock wave position shifts forward by a distance of about 97 mm. This numerical simulation indicates that the pressure recovery capacity of the supersonic separator will decline in a higher inlet pressure.

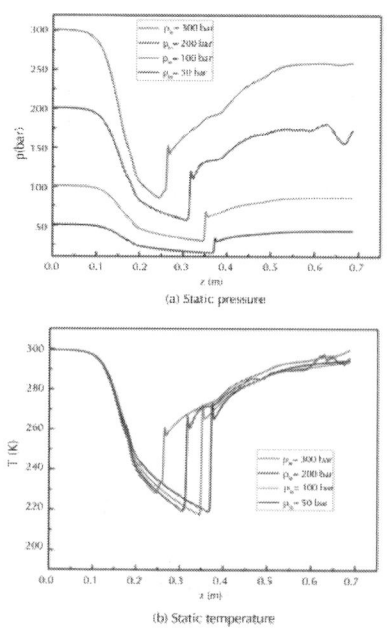

(a) Static pressure

(b) Static temperature

Figure 7: Effect of inlet pressure on natural gas dynamic parameters.

Table 4: Initial conditions for inlet pressure simulation

Cases	Inlet pressure (bar)	Inlet temperature (K)	Back pressure (bar)
1	50	300	42.5
2	100	300	85
3	200	300	170
4	300	300	255

Figure 8 depicts the phase envelope curve and the pressure–temperature (P-T) profiles with various inlet pressures. The P-T profile goes into the gas-liquid two phase zone, although the inlet pressure is changed, when the inlet pressure is lower than 100 bar. In these conditions the static pressure and temperature is low enough for the condensation of the water vapor in natural gas. But if the inlet pressure exceeds 200 bar, the natural gas flow will present a supercritical fluid in the supersonic separator, which is not suitable for the gas dehydration. Therefore, we suggest that the maximum inlet pressure should be around 100 bar for natural gas dehydration using a supersonic separator.

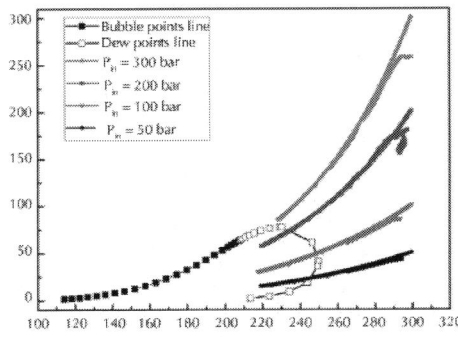

Figure 8: Phase envelope and pressure–temperature relationships with various inlet pressures.

Effect of Inlet Temperature

The inlet and back pressure are fixed to study the influence of the inlet temperature. The detailed initial conditions for inlet temperature simulation are shown in Table 5. The gas mass flow rate decreases with the rise of the inlet temperature in the supersonic separator. It indicates that the processing capacity of a supersonic separator can be improved by decreasing the inlet temperature in natural gas processing. It can be seen in Figure 9 that the shock position moves backward from nozzle to diffuser with the increase of the inlet temperature. However, the shock position moves just by a distance of about 5 mm with the increase of the inlet temperature from 10°C to 70°C, which is the normal temperature in natural gas processing. Hence, we can neglect the effect of the inlet temperature on the shock wave position in the supersonic separator. Figure 10 shows that the P-T profile goes further into the gas-liquid two phase zone with the decline of the inlet temperature. This is because the lower inlet temperature will cause a lower static temperature in the Laval nozzle, when the pressure ratio is fixed in the supersonic separator.

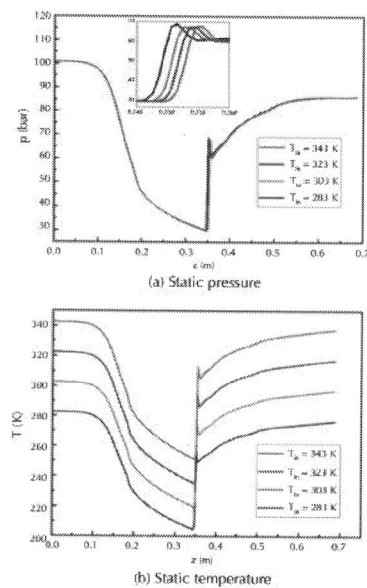

(a) Static pressure

(b) Static temperature

Figure 9: Effect of inlet temperature on natural gas dynamic parameters.

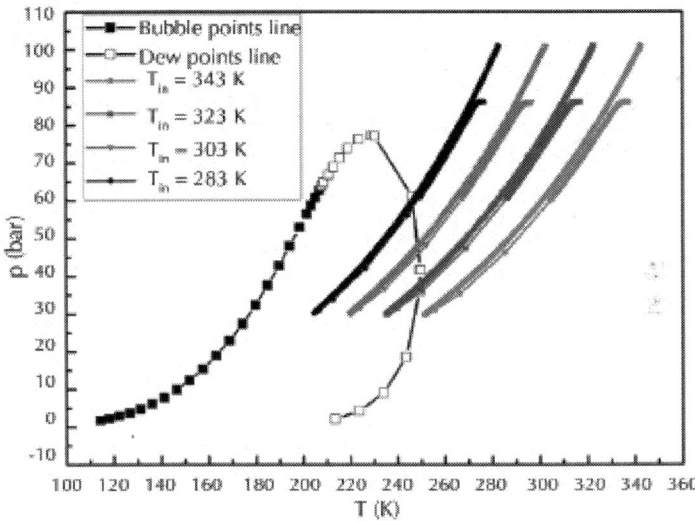

Figure 10: Phase envelope and pressure–temperature relationships with various inlet temperatures.

Table 5: Initial conditions for inlet temperature simulation

Cases	Inlet pressure (bar)	Inlet temperature (K)	Back pressure (bar)
1	100	283	85
2	100	303	85
3	100	323	85
4	100	343	85

CONCLUSIONS

The gas dynamic parameters in a supersonic separator were simulated using the Shear Stress Transport (SST) turbulence model and Redlich–

Kwong real gas model. The effect of the inlet and outlet flow conditions on the gas dynamic parameters was analyzed in the supersonic separation process, especially on the shock wave position. The gas flow cannot be choked in the supersonic separator, when the inlet mass flow rate is less than the designed one. It results in a high pressure and temperature inside the device and the water vapor cannot be removed from natural gas. The shock wave position shifts forward to the nozzle with a higher inlet pressure. The effect of the inlet temperature on the shock wave position can be neglected when the inlet temperature increases from 10°C to 70°C. The increasing back pressure induces the shock wave position to move forward from the diffuser to Laval nozzle. The shock wave moves into the supersonic channel or Laval nozzle when the back pressure is about more than 75 bar with the inlet pressure of about 100 bar. The shock wave will destroy the state of the low pressure and temperature, resulting in the re-evaporation of the condensed liquids.

ACKNOWLEDGMENTS

This work was supported in part by the National Natural Science Foundation of China (No. 51176015) and Jiangsu Key Laboratory of Oil–Gas Storage and Transportation Technology (No. SCZ1211200004/001).

AUTHOR CONTRIBUTIONS

Conceived and designed the experiments: CW YY. Performed the experiments: CW YY. Analyzed the data: CW YY. Contributed reagents/materials/analysis tools: CW YY. Wrote the paper: CW YY. Revised the manuscript: CW YY SW YF.

REFERENCES

1.   Soldo B (2012) Forecasting natural gas consumption. Appl Energy 92: 26–37. doi: 10.1016/j.apenergy.2011.11.003

2.   Economides MJ, Wood DA (2009) the state of natural gas. J Nat Gas Sci Eng 1: 1–13. doi: 10.1016/j.jngse.2009.03.005

3.   Mokhatab S, Poe WA, Speight JG (2006) Handbook of natural gas transmission and processing, Gulf Professional Publishing, Burlington, MA, USA.

4.   Okimoto D, Brouwer J (2002) Supersonic gas conditioning. World Oil 223: 89–91.

5.   Alferov VI, Baguiro LA, Dmitriev L, Feygin V, Imaev S, et al. (2005) Supersonic nozzle efficiently separates natural gas components. Oil Gas J 103: 53–58.

6.   Wen C, Cao X, Yang Y, Zhang J (2011) Swirling effects on the performance of supersonic separators for natural gas separation. Chem Eng Technol 34(9): 1575–1580. doi: 10.1002/ceat.201100095

7.   Malyshkina MM (2008) the structure of gas dynamic flow in a supersonic separator of natural gas. High Temperature 46: 69–76. doi: 10.1134/s10740-008-1010-5

8.   Malyshkina MM (2010) the procedure for investigation of the efficiency of purification of natural gases in a supersonic separator. High Temperature 48: 244–250. doi: 10.1134/s0018151x10020161

9.   Karimi A, Abdi MA (2009) Selective dehydration of high-pressure natural gas using supersonic nozzles. Chem Eng Process 48: 560–568. doi: 10.1016/j.cep.2008.09.002

10. Jiang D, Eri Q, Wang C, Tang L (2011) A fast and efficient numerical-simulation method for suprsonic gas processing. SPE projects, facilities & construction 6(2): 58–64. doi: 10.2118/131239-pa

11. Machado PB, Monteiro JGM, Medeiros JL, Epsom HD, Araujo OQF (2012) Supersonic separation in onshore natural gas dew point plant. J Nat Gas Sci Eng 6: 43–49. doi: 10.1016/j.jngse.2012.03.001

12. Mahmoodzadeh Vaziri B, Shahsavand A (2013) Analysis of supersonic separators geometry using generalized radial basis function (GRBF) artificial neural networks. J Nat Gas Sci Eng 13: 30–41. doi: 10.1016/j.jngse.2013.03.004

13. Rajaee Shooshtari SH, Shahsavand A (2013) Reliable prediction of condensation rates for purification of natural gas via supersonic separators. Sep Purif Technol 116: 458–470. doi: 10.1016/j.seppur.2013.06.009

14. Wen C, Cao X, Yang Y (2011) Swirling flow of natural gas in supersonic separators. Chem Eng Process 50(7): 644–649. doi: 10.1016/j.cep.2011.03.008

15. Wen C, Cao X, Yang Y, Li W (2012) Numerical simulation of natural gas flows in diffusers for supersonic separators. Energy 37: 195–200. doi: 10.1016/j.energy.2011.11.047

16. Wen C, Cao X, Yang Y, Zhang J (2012) Evaluation of natural gas dehydration in supersonic swirling separators applying the discrete particle method. Adv Powder Technol 23: 228–233. doi: 10.1016/j.apt.2011.02.012

17. Menter FR (1994) two-equation eddy-viscosity turbulence models for engineering applications. AIAA Journal 32(8): 1598–1605. doi: 10.2514/3.12149

18. ANSYS Fluent User Manual, 2011, ANSYS INC.

19. Kwak TY, Mansoori GA (1986) Van der waals mixing rules for cubic equations of state. Applications for supercritical fluid extraction modeling, Chem Eng Sci 41: 1303–1309. doi: 10.1016/0009-2509(86)87103-2

20. Benmekki EH, Kwak TY, Mansoori GA (1987) Supercritical fluids, American Chemical Society, Washington.

21. Redlich O, Kwong JNS (1949) On the thermodynamics of solutions. V. an equation of state. fugacities of gaseous solutions. Chem Rev 44: 233–244. doi: 10.1021/cr60137a013

22. Foelsch K (1949) The analytical design of an axially symmetric Laval nozzle for a parallel and uniform jet. J Aero Sci 16: 161–166. doi: 10.2514/8.11758

23. Patankar SV, Spalding DB (1972) A calculation procedure for heat, mass and momentum transfer in three-dimensional parabolic flows. Int J Heat Mass Transfer 1: 1787–1806. doi: 10.1016/0017-9310(72)90054-3

24. Yang Y, Wen C, Wang S, Feng Y (2014) Theoretical and numerical analysis on pressure recovery of supersonic separators for natural gas dehydration. Appl Energy 132: 248–253. doi: 10.1016/j.apenergy.2014.07.018

25. Arina R (2004) Numerical simulation of near-critical fluids. Appl Numer Math 51: 409–426. doi: 10.1016/j.apnum.2004.06.002

Citations

CHAPTER 1

Siti Syahida Mohd. Yasin, Nik Mohd. Asraf Nik Aziz, Zainal Zakaria and Ariffin Samsuri, 2014. A Study of Continuous Flow Gas Lift System Using CFD. Journal of Applied Sciences, 14: 1265-1270.

CHAPTER 2

Dabundo R, Lehmann MF, Treibergs L, Tobias CR, Altabet MA, et al. (2014) The Contamination of Commercial $^{15}N_2$ Gas Stocks with ^{15}N–Labeled Nitrate and Ammonium and Consequences for Nitrogen Fixation Measurements. PLoS ONE 9(10): e110335, doi:10.1371/journal. pone.0110335.

CHAPTER 3

Finer M, Jenkins CN, Powers B (2013) Potential of Best Practice to Reduce Impacts from Oil and Gas Projects in the Amazon. PLoS ONE 8(5): e63022. doi:10.1371/journal.pone.0063022.

CHAPTER 4

Johnson AJ, Hirson GD, Ebeler SE (2012) Perceptual Characterization and Analysis of Aroma Mixtures Using Gas Chromatography Recomposition-Olfactometry. PLoS ONE 7(8): e42693. doi:10.1371/journal.pone.0042693.

CHAPTER 5

Dressick WJ, Soto CM, Fontana J, Baker CC, Myers JD, et al. (2014) Preparation and Layer-by-Layer Solution Deposition of Cu (In,Ga)O$_2$ Nanoparticles with Conversion to Cu(In,Ga)S$_2$ Films. PLoS ONE 9(6): e100203. doi:10.1371/journal.pone.0100203.

CHAPTER 6

Valmas N, Ebert PR (2006) Comparative Toxicity of Fumigants and a Phosphine Synergist Using a Novel Containment Chamber for the Safe Generation of Concentrated Phosphine Gas. PLoS ONE 1(1): e130. doi:10.1371/journal.pone.0000130.

CHAPTER 7

Yu S, Liang Y, Sun S, Zhang K, Zhang J, et al. (2013) Vehicle Exhaust Gas Clearance by Low Temperature Plasma-Driven Nano-Titanium Dioxide Film Prepared by Radiofrequency Magnetron Sputtering. PLoS ONE 8(4): e59974. doi:10.1371/journal.pone.0059974.

CHAPTER 8

Yang Y, Wen C, Wang S, Feng Y (2014) Effect of Inlet and Outlet Flow Conditions on Natural Gas Parameters in Supersonic Separation Process. PLoS ONE 9(10): e110313. doi:10.1371/journal.pone.0110313.

Index